Lecture Notes in Computer Science 7052

Commenced Publication in 1973
Founding and Former Series Editors:
Gerhard Goos, Juris Hartmanis, and Jan van Leeuwen

Dawei Song   Massimo Melucci
Ingo Frommholz   Peng Zhang   Lei Wang
Sachi Arafat (Eds.)

Quantum Interaction

5th International Symposium, QI 2011
Aberdeen, UK, June 26-29, 2011
Revised Selected Papers

 Springer

Volume Editors

Dawei Song
Peng Zhang
Lei Wang
The Robert Gordon University
School of Computing, Aberdeen, AB25 1HG, UK
E-mail: {d.song, p.zhang1, l.wang4}@rgu.ac.uk

Massimo Melucci
University of Padua, Department of Information Engineering
Via Gradenigo, 6/B
35131 Padova, Italy
E-mail: melo@dei.unipd.it

Ingo Frommholz
University of Bedfordshire
Park Square, Luton LU1 3JU, UK
E-mail: ingo.frommholz@beds.ac.uk

Sachi Arafat
University of Glasgow, School of Computing Science
18 Lilybank Gardens, Glasgow G 128 QQ, UK
E-mail: sachi@dcs.gla.ac.uk

ISSN 0302-9743   e-ISSN 1611-3349
ISBN 978-3-642-24970-9   ISBN 978-3-642-24971-6 (eBook)
DOI 10.1007/978-3-642-24971-6
Springer Heidelberg Dordrecht London New York

Library of Congress Control Number: 2011939478

CR Subject Classification (1998): F.1, F.2.1-2, F.4.1, I.2, I.6, I.5, H.3

LNCS Sublibrary: SL 1 – Theoretical Computer Science and General Issues

Typesetting: Camera-ready by author, data conversion by Scientific Publishing Services, Chennai, India

Printed on acid-free paper

Springer is part of Springer Science+Business Media (www.springer.com)

Preface

Quantum Interaction (QI) based on Quantum Theory (QT) is being applied to domains such as artificial intelligence, human language, cognition, information retrieval, biology, political science, economics, organizations and social interaction.

After the highly successful previous meetings (QI 2007 in Stanford, QI 2008 in Oxford, QI 2009 in Saarbrücken, QI 2010 in Washington DC), the Fifth International Quantum Interaction Symposium (QI 2011) took place in Aberdeen, UK from 26 to 29 June 2011. This symposium brought together researchers interested in how QT interfaces with or solves problems in non-quantum domains more efficiently. It also looked at how QT can address previously unsolved problems in other fields.

QI 2011 received 30 submissions. All contributions were reviewed by at least three reviewers. The papers were ranked according to their relevance, originality, quality, presentation, and citations in order to decide which submissions were to be accepted as full papers, short papers, or posters. In total 11 full papers, 8 short papers and 6 posters were accepted for presentation at the conference.

These post-conference proceedings include the 23 accepted papers/posters that were presented and revised based on the reviewers' comments and the discussions at the symposium. They have been categorized into six main themes (sessions): language; semantic spaces; economics, politics and decision; psychology and cognition; information representation and retrieval; and computation and information.

We would like to thank the Steering Committee, our invited speaker Christopher Fuchs, the tutorial instructors, all the authors who submitted their work for consideration, all the participants, and the student helpers for their support and contribution; and the members of the Program Committee for their effort in providing useful and timely reviews. Our grateful thanks are also due to Ibrahim Adeyanju (local organization), Alvaro Francisco Huertas Rosero (graphical design), David Young (website design and maintenance), Steven Begg (finance), Virginia Dawood (administration), and many other people who offered great help. We also would like to acknowledge the financial support from the Scottish Informatics and Computer Science Alliance (SICSA).

Finally, we hope everybody had a fruitful and enjoyable time in Aberdeen.

July 2011

Dawei Song
Massimo Melucci
Ingo Frommholz
Peng Zhang
Lei Wang
Sachi Arafat

Organization

Program Committee

Diederik Aerts	Free University Brussels
Sven Aerts	Free University Brussels
Sachi Arafat	University of Glasgow
Harald Atmanspacher	Institute for Frontier Areas of Psychology and Mental Health (IGPP)
Peter Bruza	Queensland University of Technology
Jerome Busemeyer	Indiana University
Bob Coecke	Oxford University
Trevor Cohen	University of Texas, Houston
Riccardo Franco	Politecnico di Torino
Ingo Frommholz	University of Bedfordshire
Liane Gabora	University of British Columbia
Emmanuel Haven	University of Leicester
Andre Khrennikov	Linnaeus University
Kirsty Kitto	Queensland University of Technology
Ariane Lambert-Mogiliansky	Paris School of Economics
William Lawless	Paine College
Massimo Melucci	University of Padua
Jian-Yun Nie	Université de Montréal
Dusko Pavlovic	Kestrel Institute and Oxford University
Don Sofge	Naval Research Laboratory
Dawei Song	The Robert Gordon University
Keith van Rijsbergen	University of Glasgow
Salvador Venegas-Andraca	Tecnológico de Monterrey
Giuseppe Vitiello	University of Salerno
Jun Wang	The Robert Gordon University
Dominic Widdows	Google Inc.
John Woods	University of British Columbia
Mingsheng Ying	University of Technology Sydney
Vyacheslav Yukalov	Joint Institute for Nuclear Research

Additional Reviewers

De Vine, Lance
Haven, Emmanuel
Veloz, Tomas

Table of Contents

Keynote Talk

Born's Rule as an Empirical Addition to Probabilistic Coherence....... 1
  Christopher A. Fuchs

Language

Introducing Scalable Quantum Approaches in Language
Representation 2
  Peter Wittek and Sándor Darányi

Similarity Metrics within a Point of View 13
  Sven Aerts, Kirsty Kitto, and Laurianne Sitbon

Toward a Formal Model of the Shifting Relationship between Concepts
and Contexts during Associative Thought 25
  Tomas Veloz, Liane Gabora, Mark Eyjolfson, and Diederik Aerts

Semantic Spaces

A Compositional Distributional Semantics, Two Concrete
Constructions, and Some Experimental Evaluations 35
  Mehrnoosh Sadrzadeh and Edward Grefenstette

Finding Schizophrenia's Prozac: Emergent Relational Similarity in
Predication Space... 48
  *Trevor Cohen, Dominic Widdows, Roger Schvaneveldt, and
  Thomas C. Rindflesch*

Spectral Composition of Semantic Spaces 60
  Peter Wittek and Sándor Darányi

Economics, Politics and Decision

Dynamic Optimization with Type Indeterminate Decision-Maker:
A Theory of Multiple-self Management 71
  Ariane Lambert-Mogiliansky and Jerome Busemeyer

Pseudo-classical Nonseparability and Mass Politics in Two-Party
Systems 83
  Christopher Zorn and Charles E. Smith Jr.

A Quantum Cognition Analysis of the Ellsberg Paradox 95
   Diederik Aerts, Bart D'Hooghe, and Sandro Sozzo

Psychology and Cognition

Can Classical Epistemic States Be Entangled? 105
   Harald Atmanspacher, Peter beim Graben, and Thomas Filk

Quantum Structure in Cognition: Why and How Concepts Are
Entangled.. 116
   Diederik Aerts and Sandro Sozzo

Options for Testing Temporal Bell Inequalities for Mental Systems 128
   Harald Atmanspacher and Thomas Filk

Information Representation and Retrieval

Quantum-Like Uncertain Conditionals for Text Analysis 138
   Alvaro Francisco Huertas-Rosero and C.J. van Rijsbergen

Modelling the Acitivation of Words in Human Memory: The Spreading
Activation, Spooky-Activation-at-a-Distance and the Entanglement
Models Compared... 149
   *David Galea, Peter Bruza, Kirsty Kitto, Douglas Nelson, and
   Cathy McEvoy*

Senses in Which Quantum Theory Is an Analogy for Information
Retrieval and Science 161
   Sachi Arafat

Computation and Information

A Hierarchical Sorting Oracle.................................... 172
   Luís Tarrataca and Andreas Wichert

Quantum-Like Paradigm: From Molecular Biology to Cognitive
Psychology... 182
   *Masanari Asano, Masanori Ohya, Yoshiharu Tanaka,
   Ichiro Yamato, Irina Basieva, and Andrei Khrennikov*

Posters

A Quantum-Conceptual Explanation of Violations of Expected Utility
in Economics 192
   Diederik Aerts, Jan Broekaert, Marek Czachor, and Bart D'Hooghe

On the Nature of the Human Mind: The Cognit Space Theory   199
George Economides

Quantum Phenomenology and Dynamic Co-Emergence   205
Christian Flender

Envisioning Dynamic Quantum Clustering in Information Retrieval   211
Emanuele Di Buccio and Giorgio Maria Di Nunzio

Contextual Image Annotation via Projection and Quantum Theory
Inspired Measurement for Integration of Text and Visual Features   217
Leszek Kaliciak, Jun Wang, Dawei Song, Peng Zhang, and
Yuexian Hou

MPEG-7 Features in Hilbert Spaces: Querying Similar Images with
Linear Superpositions   223
Elisa Maria Todarello, Walter Allasia, and Mario Stroppiana

Author Index   229

Born's Rule as an Empirical Addition to Probabilistic Coherence

Christopher A. Fuchs

Perimeter Institute for Theoretical Physics
Waterloo, Ontario
Canada
cfuchs@perimeterinstitute.ca

Abstract. With the help of a certain mathematical structure in quantum information theory, there is a particularly elegant way to rewrite the quantum mechanical Born rule as an expression purely in terms of probabilities.

In this way, one can in principle get rid of complex Hilbert spaces and operators as fundamental entities in the theory. In the place of a quantum state, the new expression uses a probability distribution, and in the place of measurement operators, it uses conditional distributions.

The Born rule thus becomes a story of probabilities going in and probabilities coming out. Going a step further: In the Bayesian spirit of giving equal status to all probabilities – in this case, the ones on both the right and left sides of the Born-rule equation – it indicates that the Born rule should be viewed as a normative condition on probabilities above and beyond Dutch-book coherence.

In opposition to Dutch book coherence, this new normative rule is empirical, rather than purely logical in its origin (and by way of that must encode some of the physical content of quantum theory), but there may be other non-quantum situations that warrant the same or a similar addition to Dutch-book coherence: I make no judgment one way or the other, but I hope that this way of rewriting quantum theory may provide a suggestive new language for some of the non-quantum topics of this meeting.

D. Song et al. (Eds.): QI 2011, LNCS 7052, p. 1, 2011.

Introducing Scalable Quantum Approaches in Language Representation

Peter Wittek and Sándor Darányi

Swedish School of Library and Information Science
Göteborg University & University of Borås
Allégatan 1, 50190 Borås, Sweden
peterwittek@acm.org, sandor.daranyi@hb.se

Abstract. High-performance computational resources and distributed systems are crucial for the success of real-world language technology applications. The novel paradigm of general-purpose computing on graphics processors (GPGPU) offers a feasible and economical alternative: it has already become a common phenomenon in scientific computation, with many algorithms adapted to the new paradigm. However, applications in language technology do not readily adapt to this approach. Recent advances show the applicability of quantum metaphors in language representation, and many algorithms in quantum mechanics have already been adapted to GPGPU computing. SQUALAR aims to match quantum algorithms with heterogeneous computing to develop new formalisms of information representation for natural language processing in quantum environments.

1 Introduction

Quantum mechanics is a very successful scientific theory for making predictions about systems with inherent ambiguity in them. That natural language bears similarities with such a system is at least plausible. Recent advances in theory and experimentation to apply quantum mechanics to non-quantum domains include the use of quantum algorithms to address, or to more efficiently solve, problems in such domains (including contrasts between classical vs. quantum methods), such as applications of artificial intelligence, information retrieval, and language modelling.

The quantum metaphor promises improved methodologies to capture the subtleties and ambiguities of human language, resulting in optimised algorithms for text processing. The purpose of SQUALAR is to investigate methods borrowed from the field of quantum mechanics in a wide range of large-scale language technology applications by seeking a match between quantum algorithms and heterogeneous computing.

To this end, a scalable environment is a must. Latest trends indicate the rise of a heterogeneous platform in which multi-core central processing units (CPUs) and graphics processing units (GPUs) work together in a distributed-memory parallelism. CPU-based parallelism has been utilized for decades, and while not

D. Song et al. (Eds.): QI 2011, LNCS 7052, pp. 2–12, 2011.

without its own problems, it is a mature field and multicore CPUs enable developing faster algorithms with reasonable effort. In this paradigm, there is a considerable overhead on dividing the problem, distributing the bits along a small number of CPU cores, then collecting and merging results. This type of parallelism is available in a wide range of programming languages, although the source code needs to be modified to some extent. GPU-based parallelism is a completely different approach. The overhead of splitting the work is minimal, the number of cores is massive, but the kind of computations that can be split is limited to a simple, single-pass operation. This heterogeneous computing environment has to be studied at different levels to find scalable implementations: low-level linear algebra, numerical methods, kernel methods and manifold learning are candidates for testing, as well as higher level load distribution such as MapReduce [1]. The constraints are as follows:

- Text processing is typically a data-intensive task, and several distributed algorithms have been proposed to deal with large-scale collections on a grid or in a cloud computing environment. MapReduce[1] was originally developed to this end, and mature libraries, such as Cloud9, are readily available [2]. Other libraries, such as Mahout[2], facilitate the development of complex language technology applications.
- General-purpose computing on the GPU requires considerable effort from developers. Initial results in text processing, however, indicate that the improvement in execution time can be considerable [3–7].
- Quantum methods, on the other hand, rely on linear algebra and other numerical libraries, many of which have already been optimized to utilize the power of GPUs [8–11].

SQUALAR intends to bring the best of two worlds together. By bridging data-intensive text processing with sophisticated quantum modelling of languages, we expect to see major advances in language technology.

The challenges, however, are far from trivial. The major frameworks of GPGPU programming, CUDA and OpenCL, require wrapping in Java, which is the environment of Hadoop, the most mature open source MapReduce implementation. This paper offers an insight on the initial stage of our ongoing investigation.

This paper is organized as follows. Section 2 defines what we mean by heterogeneous computing: a distributed system of nodes which are equipped with multicore CPUs and GPUs. Section 3 gives a very short overview of quantum approaches in language processing, with a focus on methods that have the potential for acceleration. Section 4 discusses how we intend to bridge heterogeneous computing and these quantum approaches, and finally Section 5 concludes our paper.

---

[1] http://hadoop.apache.org/mapreduce/
[2] http://mahout.apache.org

2   Heterogeneous Computing

Heterogeneous computing aims to combine the parallelism of traditional multicore CPUs and GPU accelerator cores to deliver unprecedented levels of performance [12]. While the phrase typically refers to single node, a distributed environment may be constructed from such heterogeneous nodes.

CPUs excel in running single-threaded processes, or in multithreaded applications in which a thread often consists of fairly complicated sequential code. Graphics processors are ideally suited for computations that can be run on numerous data elements simultaneously in parallel. This typically involves arithmetic on large data sets (such as matrices) where the same operation can be performed across thousands of elements at the same time. This is actually a requirement for good performance: the software must use a large number of threads. The overhead of creating new threads is minimal compared to CPUs that typically take thousands of clock cycles to generate and schedule, and a low number of threads will not perform well on GPU [13]. The decomposition and scheduling of computation among CPU cores and GPUs are not trivial even on a single node [14–16], and the task is even more complicated for clusters [17]. In order to issue work to several GPUs concurrently, a program needs the same number of CPU threads, each with its own context. All inter-GPU communication takes place via host nodes. Threads can be lightweight (pthreads, OpenMP, etc. [18]) or heavyweight (MPI [19]). Any CPU multi-threading or message-passing API or library can be used, as CPU thread management is completely orthogonal to GPGPU programming. For example, one can add GPU processing to an existing MPI application by porting the compute-intensive portions of the code without changing the communication structure [20]. However, the efficient utilisation of all CPU and GPU cores remains an open question.

While research is being carried out to develop the formal foundations of efficient scheduling and decomposition in multiple heterogeneous nodes, GPU-based clouds are becoming available[3,4], and initial investigations have been carried out to develop an efficient MapReduce framework [21, 22]. Like OpenMP and MPI, MapReduce provides an abstraction, a means to distribute computation without burdening the programmer with the details of distributed computing; however, the level of granularity is different [2]. These frameworks are mostly designed to deal with processor-intensive problems and have only rudimentary support for dealing with very large amounts of input data. The strength of MapReduce is data-intensive distributed parallel processing on a massive scale [1]. The potential of combining a data-intensive cloud-based approach with the compute-intensive GPGPU paradigm for sophisticated, large-scale natural language processing is enormous.

---

[3] http://www.hoopoe-cloud.com/
[4] http://aws.typepad.com/aws/2010/11/new-ec2-instance
-type-the-cluster-gpu-instance.html

3   Quantum Approaches in Language Processing

Metaphors of quantum theory in linguistic applications arose over the last decade [23–28]. The vector space model of information retrieval was first adopted largely because it allowed for a naturally continuous 'relevance score' by using the cosine dissimilarity, as opposed a mere binary decision between relevant and irrelevant documents. In a similar fashion quantum mechanics yields a continuous probability that a particular event will be observed, a feature making it useful to reflect on possible similarities with natural language [24]. Moreover, it appears likely that quantum interaction would be of a type where the context of the interaction itself must be incorporated into the model. For example, a measurement in a quantum-scale system will have an impact on the result. If the system is displaying contextual behaviour such as natural languages, then a quantum approach often incorporates this behaviour very naturally [29].

Quantum phenomena in languages may be present at different levels. At subword level, terms and documents can be regarded as linear combinations of their semantic features [30], which can account for semantic priming [31].

At word level, a word in semantic space may be likened to a quantum particle. In the absence of context it is in a superposed state, it is a collection of all the possible meanings of the word: $\rho = p_1\rho_1 + \ldots + p_m\rho_m$, where ρ is the word in the semantic space as a density matrix, and each i is a basis state representing one of the m senses of the word and the probabilities p_i sum to unity. Encountering the word in context, however, gives rise to a 'collapse' of potential meanings onto an actual one. The context is modelled a projection operator which is applied to a given density matrix corresponding to the state of a word meaning resulting in its 'collapse' [32].

Turning to combinations of words, at least two approaches offer solutions. One uses the operator algebra of quantum theory to construct a 'semantic calculus' [26, 33]. The other approach encodes word order relying on random indexing [34, 35], using either permutation [36, 37] or circular convolution [31, 37]. The order can also be encoded by tensor product [25, 38].

Using different units of analysis, quantum approaches find their way to applications, most notably:

- Information retrieval: Vector space logic and quantum logic (Neumann algebra) are very similar [26]. In particular, negation has been investigated in depth in [24]. These models may allow new types of queries and also inference [39].
- Memory models: Two schools of memory models are relevant to quantum theory: matrix memory [40, 41, 32], and convolution-correlation memory (holographic-like) [42, 43, 31]. Matrix models are not directly related to QT, but there can be a connection through Heisenberg's matrix mechanics, which was the first complete and correct definition of quantum mechanics. It is equivalent to the Schrödinger wave formulation of quantum mechanics, and is the basis of Dirac's bra-ket notation for the wave function. Matrix models can incorporate hierarchical sentence and paragraph representation

[44], bridging distributional and symbolic approaches [45], encode location [46], or include sense in a term-space approach [47, 48]. The other approach, convolution memory is particularly useful to encode syntactic information [49, 37].

- Semantic regions: Regions meant to solve the problem to be able to say that apple is a kind of fruit (apple is part of the fruit region), as opposed to modelling that apples and fruit have something to do with one another. Separating hyperspaces may define a semantic region [33]. As an alternative, [50] measures the distance between subspaces spanned by documents by projecting them into one another.
- Spectral theory in mathematics is key to the success of as diverse application domains as quantum mechanics and methods of latent semantic analysis (LSA, [51]) for language processing. In other words, both quantum mechanics and LSA rely on eigenvalue decomposition for the localization of their respective entities in observation space. This neglected fact, supported by a high number of papers in different disciplines describing the dynamic behaviour of documents and their index terms over time, points at some implicit "energy" inherent in them and in need of quantification. Prominently, theories of word meaning (contextual [52, 53] and referential [54, 55]), currently used in applications trying to capture and exploit semantic content, fall back on quantities of qualities, but quite possibly miss the underlying framework. LSA is just one spectral approach in language representation: [32] demonstrate the quantum collapse of meaning using the hyperspace analogue to language (HAL, [56]).

4   Methods and Planned Outcomes

With the above plethora of approaches available for testing, the fundamental task of SQUALAR is bridging scalable linear algebra and numerical methods that are widely used in scientific computing with the emerging theories in quantum interaction to enable practical, real-world language technology applications.

The hardware and basic software infrastructure is what we described in section 2: a distributed system consisting of heterogeneous nodes which combine multicore CPUs and GPUs (top part of Figure 1). Since hardware virtualization is already at consumer level, the distributed system can be either a privately owned cluster or grid, or a high-performance computing cloud provided by a third-party.

Without going into details, algorithms in linear algebra are the most obvious candidates for acceleration on graphics hardware (middle part of Figure 1, left). Vector space models of semantics can be implemented by accelerated BLAS libraries [8, 10], including operator algebra for semantic inference [24, 26]. Matrix decompositions and dimension reduction that also play an important role in understanding semantics are currently limited to matrices of limited sizes [11]. Convolution, which plays an important part in encoding term positions [31, 37], can be mapped to the frequency domain by Fourier transformation, where

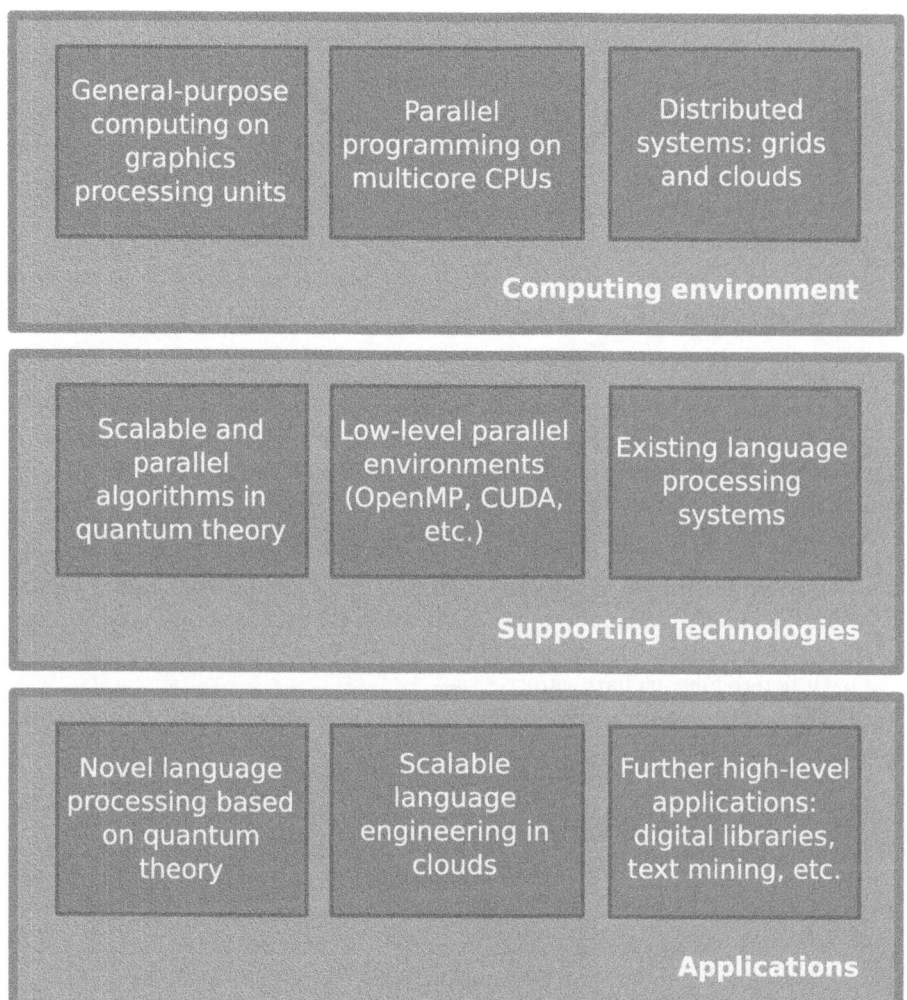

Fig. 1. An overview of the SQUALAR framework

the operation simplifies to a simple multiplication. Fast Fourier transformation on GPUs is a classical area for acceleration [57]. More complex examples in accelerated quantum methods [58, 59] and related visualization [60] are awaiting appropriate metaphors in language processing.

Approaching from existing language processing algorithms, if a sufficient metaphor cannot be found or if the method does not lend itself easily to any of the methods described above, lower level libraries can be used for developing multithreaded, GPU-based implementations (middle part of Figure 1, right and middle).

If we focus on a single computer, we will be able to perform operations several folds faster, gaining new insights on language technology (bottom part of Figure

1, left). By providing a high-level load balancing mechanism, the potential of compute and data-intensive processing can be released in a distributed environment for web-scale applications (bottom part of Figure 1, middle). Some machine learning algorithms, such as support vector machines, have already been adopted to graphics hardware [61]. Combining these with the above, we gain powerful text mining applications (bottom part of Figure 1, right). Since Information Retrieval has already began experimenting with a wide range of quantum theory based metaphors, this field has the most to benefit.

5   Conclusion

Whether language to some extent shares a conceptual framework with quantum mechanics, and if thereby some linguistic phenomena could be eventually modelled on physical ones, is a research question of interest to SQUALAR. We trust that by better mastering the match between quantum algorithms and GPU computing, web-scale applications will become feasible.

The fundamental tasks and challenges of the project are the following:

- Rephrasing natural language processing and text mining algorithms in quantum domain to use compute-intensive heterogeneous programming model;
- Data and compute-intensive distributed and cloud computing applications with heterogeneous hardware;
- Performance evaluation of heterogeneous hardware for natural language processing tasks;
- Trade-offs of using scalable quantum models in language engineering;
- Exploiting heterogeneous architectures to accelerate sophisticated language processing.

Acknowledgement. We would like to thank Lance de Vine (Queensland University of Technology) for discussions related to ideas presented in this paper. This work was also supported by Amazon Web Services.

References

1. Dean, J., Ghemawat, S.: MapReduce: Simplified data processing on large clusters. In: Proceedings of OSDI 2004, 6th International Symposium on Operating Systems Design & Implementation, San Francisco, CA, USA. ACM Press, New York (2004)
2. Lin, J., Dyer, C.: Data-Intensive Text Processing with MapReduce. Morgan & Claypool (2010)
3. Cavanagh, J., Potok, T., Cui, X.: Parallel latent semantic analysis using a graphics processing unit. In: Proceedings of GECCO 2009, 11th Annual Conference Companion on Genetic and Evolutionary Computation Conference: Late Breaking Papers, Montreal, QC, Canada, pp. 2505–2510. ACM Press, New York (2009)
4. Ding, S., He, J., Yan, H., Suel, T.: Using graphics processors for high performance IR query processing. In: Proceedings of WWW 2009, 18th International Conference on World Wide Web, Spain, Madrid, pp. 421–430. ACM Press, New York (2009)

5. Zhang, Y., Mueller, F., Cui, X., Potok, T.: Large-scale multi-dimensional document clustering on GPU clusters. In: Proceedings of IDPDS 2010, 24th International Parallel and Distributed Computing Symposium, Atlanta, GA, USA. IEEE Computer Society Press, Los Alamitos (2010)

6. Byna, S., Meng, J., Raghunathan, A., Chakradhar, S., Cadambi, S.: Best-effort semantic document search on GPUs. In: Proceedings of GPGPU 2010, 3rd Workshop on General-Purpose Computation on Graphics Processing Units, pp. 86–93. ACM, New York (2010)

7. Wei, Z., JaJa, J.: A fast algorithm for constructing inverted files on heterogeneous platforms. In: Proceedings of IPDPS 2011, 25th International Parallel and Distributed Computing Symposium, Anchorage, AK, USA (2011)

8. Krüger, J., Westermann, R.: Linear algebra operators for GPU implementation of numerical algorithms. In: Proceedings of SIGGRAPH 2005, 32nd International Conference on Computer Graphics and Interactive Techniques, Los Angeles, CA, USA, pp. 234–242. ACM Press, New York (2005)

9. Galoppo, N., Govindaraju, N., Henson, M., Bondhugula, V., Larsen, S., Manocha, D.: Efficient numerical algorithms on graphics hardware. In: Proceedings of EDGE 2006, Workshop on Edge Computing Using New Commodity Architectures, Chapel Hill, NC, USA (2006)

10. Barrachina, S., Castillo, M., Igual, F., Mayo, R., Quintana-Orti, E.: Evaluation and tuning of the level 3 CUBLAS for graphics processors. In: Proceedings of IPDPS 2008, 22nd International Symposium on Parallel and Distributed Processing, Miami, FL, USA, pp. 1–8. IEEE, Los Alamitos (2008)

11. Lahabar, S., Narayanan, P.: Singular value decomposition on GPU using CUDA. In: Proceedings of IPDPS 2009, 23rd International Symposium on Parallel and Distributed Processing, Rome, Italy, IEEE, Los Alamitos (2009)

12. Brodtkorb, A., Dyken, C., Hagen, T., Hjelmervik, J., Storaasli, O.: State-of-the-art in heterogeneous computing. Scientific Programming 18(1), 1–33 (2010)

13. Kirk, D., Hwu, W.: Programming massively parallel processors: A hands-on approach (2009)

14. Jiménez, V., Vilanova, L., Gelado, I., Gil, M., Fursin, G., Navarro, N.: Predictive runtime code scheduling for heterogeneous architectures. High Performance Embedded Architectures and Compilers, 19–33 (2009)

15. Lee, S., Min, S.J., Eigenmann, R.: OpenMP to GPGPU: a compiler framework for automatic translation and optimization. In: Proceedings of PPOPP 2009, 14th Symposium on Principles and Practice of Parallel Programming, pp. 101–110. ACM Press, New York (2009)

16. Luk, C., Hong, S., Kim, H.: Qilin: Exploiting parallelism on heterogeneous multiprocessors with adaptive mapping. In: MICRO-42, 42nd Annual IEEE/ACM International Symposium on Microarchitecture, New York, NY, USA, pp. 45–55. IEEE, Los Alamitos (2009)

17. Phillips, J., Stone, J., Schulten, K.: Adapting a message-driven parallel application to GPU-accelerated clusters. In: Proceedings of SC 2008, 21st Conference on Supercomputing, Austin, TX, USA, pp. 1–9. IEEE Press, Los Alamitos (2008)

18. Kuhn, B., Petersen, P., O'Toole, E.: OpenMP versus threading in C/C++. Concurrency: Practice and Experience 12(12), 1165–1176 (2000)

19. Koop, M., Sur, S., Gao, Q., Panda, D.: High performance MPI design using unreliable datagram for ultra-scale InfiniBand clusters. In: Proceedings of ISC-06, 21st Annual International Conference on Supercomputing, Dresden, Germany, pp. 180–189. ACM, New York (2006)

20. NVida Compute Unified Device Architecture Best Practices Guide 3.2 (2010)
21. Shirahata, K., Sato, H., Matsuoka, S.: Hybrid map task scheduling on GPU-based heterogeneous clusters. In: Proceedings of CloudCom 2010, The 2nd International Conference on Cloud Computing, Indianapolis, IN, USA (2010)
22. Stuart, J., Owens, J.: Multi-GPU MapReduce on GPU clusters. In: Proceedings of IPDPS 2011, 25th International Parallel and Distributed Computing Symposium, Anchorage, AK, USA (2011)
23. Aerts, D., Aerts, S., Broekaert, J., Gabora, L.: The violation of bell inequalities in the macroworld. Foundations of Physics 30(9), 1387–1414 (2000)
24. Widdows, D., Peters, S.: Word vectors and quantum logic: Experiments with negation and disjunction. In: Proceedings of MoL 2003, 8th Mathematics of Language Conference, Bloomington, IN, USA, vol. 8, pp. 141–154 (2003)
25. Aerts, D., Czachor, M.: Quantum aspects of semantic analysis and symbolic artificial intelligence. Journal of Physics A: Mathematical and General 37, L123–L132 (2004)
26. van Rijsbergen, C.J.: The Geometry of Information Retrieval. Cambridge University Press, New York (2004)
27. Widdows, D.: Geometry and meaning (2004)
28. Bruza, P., Widdows, D., Woods, J.: A quantum logic of down below. In: Engesser, K., Gabbay, D., Lehmann, D. (eds.) Handbook of Quantum Logic and Quantum Structures, vol. 2, Elsevier, Amsterdam (2009)
29. Kitto, K.: Why quantum theory? In: Proceedings of QI 2008, 2nd International Symposium on Quantum Interaction, Oxford, UK, pp. 11–18 (2008)
30. Lyons, J.: Semantics. Cambridge University Press, New York (1977)
31. Jones, M., Mewhort, D.: Representing word meaning and order information in a composite holographic lexicon. Psychological Review 114(1), 1–37 (2007)
32. Bruza, P., Woods, J.: Quantum collapse in semantic space: interpreting natural language argumentation. In: Proceedings of QI 2008, 2nd International Symposium on Quantum Interaction, Oxford, UK. College Publications (2008)
33. Widdows, D.: Semantic vector products: Some initial investigations. In: Proceedings of QI 2008, 2nd International Symposium on Quantum Interaction. College Publications, Oxford (2008)
34. Kanerva, P., Kristofersson, J., Holst, A.: Random indexing of text samples for latent semantic analysis. In: Proceedings of CogSci 2000, 22nd Annual Conference of the Cognitive Science Society, Philadelphia, PA, USA, vol. 1036 (2000)
35. Sahlgren, M.: An introduction to random indexing. In: Proceedings of TKE 2005, Methods and Applications of Semantic Indexing Workshop at the 7th International Conference on Terminology and Knowledge Engineering, Copenhagen, Denmark, Citeseer (2005)
36. Sahlgren, M., Holst, A., Kanerva, P.: Permutations as a means to encode order in word space. In: Proceedings of CogSci 2008, 30th Annual Meeting of the Cognitive Science Society, Washington, DC, USA (2008)
37. De Vine, L., Bruza, P.: Semantic oscillations: Encoding context and structure in complex valued holographic vectors. In: Proceedings of QI 2010, 4th Symposium on Quantum Informatics for Cognitive, Social, and Semantic Processes, Arlington, VA, USA, pp. 11–13 (2010)
38. Mitchell, J., Lapata, M.: Vector-based models of semantic composition. In: Proceedings of ACL 2008, 46th Annual Meeting of the Association for Computational Linguistics, Columbus, Ohio, pp. 236–244. ACL, Morristown (2008)

39. Song, D., Lalmas, M., van Rijsbergen, C., Frommholz, I., Piwowarski, B., Wang, J., Zhang, P., Zuccon, G., Bruza, P., Arafat, S., et al.: How quantum theory is developing the field of Information Retrieval. In: Proceedings of QI 2010, 4th Symposium on Quantum Informatics for Cognitive, Social, and Semantic Processes, Arlington, VA, USA, pp. 105–108 (2010)
40. Humphreys, M., Bain, J., Pike, R.: Different ways to cue a coherent memory system: A theory for episodic, semantic, and procedural tasks. Psychological Review 96(2), 208–233 (1989)
41. Wiles, J., Halford, G., Stewart, J., Humphreys, M., Bain, J., Wilson, W.: Tensor models: A creative basis for memory retrieval and analogical mapping. In: Dartnall, T. (ed.) Artificial Intelligence and Creativity, pp. 145–159. Kluwer Academic, Dordrecht (1994)
42. Plate, T.: Holographic reduced representations: Convolution algebra for compositional distributed representations. In: Proceedings of IJCAI 1991, 12th International Joint Conference on Artificial Intelligence, Syndey, Australia, Citeseer, pp. 30–35 (1991)
43. Plate, T.: Holographic reduced representations. IEEE Transactions on Neural Networks 6(3), 623–641 (1995)
44. Antonellis, I., Gallopoulos, E.: Exploring term-document matrices from matrix models in text mining. In: Proceedings of SDM 2006, Text Mining Workshop in Conjuction with the 6th SIAM International Conference on Data Mining, Bethesda, MD, USA (2006)
45. Rudolph, S., Giesbrecht, E.: Compositional matrix-space models of language. In: Proceedings of ACL 2010, 48th Annual Meeting of the Association for Computational Linguistics, Uppsala, Sweden, pp. 907–916. Association for Computational Linguistics (2010)
46. Rölleke, T., Tsikrika, T., Kazai, G.: A general matrix framework for modelling information retrieval. Information Processing & Management 42(1), 4–30 (2006)
47. Swen, B.: A sense matrix model for information retrieval. Technical report TR-2004-2 of ICL-PK (2004)
48. Novakovitch, D., Bruza, P., Sitbon, L.: Inducing shades of meaning by matrix methods: a first step towards thematic analysis of opinion. In: Proceedings of SEMAPRO 2009, 3rd International Conference on Advances in Semantic Processing, Sliema, Malta, pp. 86–91. IEEE, Los Alamitos (2009)
49. Jones, M., Kintsch, W., Mewhort, D.: High-dimensional semantic space accounts of priming. Journal of Memory and Language 55(4), 534–552 (2006)
50. Zuccon, G., Azzopardi, L.A., van Rijsbergen, C.J.: Semantic spaces: Measuring the distance between different subspaces. In: Bruza, P., Sofge, D., Lawless, W., van Rijsbergen, K., Klusch, M. (eds.) QI 2009. LNCS, vol. 5494, pp. 225–236. Springer, Heidelberg (2009)
51. Deerwester, S., Dumais, S., Furnas, G., Landauer, T., Harshman, R.: Indexing by latent semantic analysis. Journal of the American Society for Information Science 41(6), 391–407 (1990)
52. Wittgenstein, L.: Philosophical Investigations. Blackwell Publishing, Oxford (1967)
53. Harris, Z.: Distributional structure. In: Harris, Z. (ed.) Papers in Structural and Transformational Linguistics. Formal Linguistics, pp. 775–794. Humanities Press, New York (1970)
54. Peirce, C.: Logic as semiotic: The theory of signs. In: Peirce, C.S., Buchler, J. (eds.) Philosophical Writings of Peirce, pp. 98–119. Dover Publications, Mineola (1955)
55. Frege, G.: Sense and reference. The Philosophical Review 57(3), 209–230 (1948)

56. Lund, K., Burgess, C.: Producing high-dimensional semantic spaces from lexical co-occurrence. Behavior Research Methods Instruments and Computers 28, 203–208 (1996)
57. Govindaraju, N., Lloyd, B., Dotsenko, Y., Smith, B., Manferdelli, J.: High performance discrete Fourier transforms on graphics processors. In: Proceedings of SC 2008, 21st Conference on Supercomputing, Austin, TX, USA. IEEE, Los Alamitos (2008)
58. Ufimtsev, I., Martínez, T.: Graphical processing units for quantum chemistry. Computing in Science & Engineering 10(6), 26–34 (2008)
59. Watson, M., Olivares-Amaya, R., Edgar, R., Aspuru-Guzik, A.: Accelerating correlated quantum chemistry calculations using graphical processing units. Computing in Science & Engineering 12(4), 40–51 (2010)
60. Stone, J., Saam, J., Hardy, D., Vandivort, K., Hwu, W., Schulten, K.: High performance computation and interactive display of molecular orbitals on GPUs and multi-core CPUs. In: Proceedings of GPGPU 2009, 2nd Workshop on General Purpose Processing on Graphics Processing Units, Washington, DC, USA, pp. 9–18. ACM, New York (2009)
61. Catanzaro, B., Sundaram, N., Keutzer, K.: Fast support vector machine training and classification on graphics processors. In: McCallum, A., Roweis, S. (eds.) Proceedings of ICML 2008, 25th Annual International Conference on Machine Learning, Helsinki, Finland, pp. 104–111. Omnipress (2008)

Similarity Metrics within a Point of View

Sven Aerts[1], Kirsty Kitto[2], and Laurianne Sitbon[2]

[1] Centre for Interdisciplinary Studies CLEA,
Vrije Universiteit Brussel
saerts@vub.ac.be
[2] Faculty of Science and Technology,
Queensland University of Technology
{kirsty.kitto,laurianne.sitbon}@qut.edu.au

Abstract. Vector space based approaches to natural language processing are contrasted with human similarity judgements to show the manner in which human subjects fail to produce data which satisfies all requirements for a metric space. This result would constrains the validity and applicability vector space based (and hence also quantum inspired) approaches to the modelling of cognitive processes. This paper proposes a resolution to this problem, by arguing that pairs of words imply a context which in turn induces a point of view, so allowing a subject to estimate semantic similarity. Context is here introduced as a point of view vector (POVV) and the expected similarity is derived as a measure over the POVV's. Different pairs of words will invoke different contexts and different POVV's. We illustrate the proposal on a few triples of words and outline further research.

Keywords: Similarity, Semantic Space, Triangle Inequality, Metric, Context.

1 Introduction

Human language is frequently represented in a mental lexicon, which refers to both the words in that language, and its structure, or the set of associative links which bind this vocabulary together. Such links are acquired through experience, and the vast and semi-random nature of this experience ensures that words within this vocabulary are highly interconnected, both directly and indirectly through other words. For example, the word *planet* can become associated with *earth*, *space*, *moon*, and so on, and within this set, *moon* can become linked to *earth* and *star* [7].

The complexity of the mental lexicon makes it challenging to construct analytical and computational models of both its structure and behavior. Yet even relatively small steps towards achieving the automatic interpretation of human language have given us search engines capable of converting our human made queries into their mathematical equivalent, and identifying documents relevant to that query among the huge corpus of the internet. Thus, these small steps have transformed the way we use the internet today. It seems clear that having a

D. Song et al. (Eds.): QI 2011, LNCS 7052, pp. 13–24, 2011.

better mathematical representation of human language will lead to an improved use of the information content of the internet, however, the question of how to best represent human language remains a theoretical challenge. In this paper we shall consider one particular challenge, that of metricity. While vector space based models of the human mental lexicon have proven successful in various respects, the manner in which they quantize similarity is different from human judgements of semantic similarity, which violate key properties required of a metric[24]. We shall then propose a contextual resolution to this problem and conclude by suggesting some potential future avenues of investigation. We begin with a brief overview of current vector space models of the mental lexicon.

2   Vector Space Models of the Mental Lexicon

Computational representations of the mental lexicon have been investigated by researchers from a range of disciplines, including mathematics, logic, philosophy, artificial intelligence, computational linguistics, cognitive psychology, natural language processing and information retrieval [23]. The birth of vector space based models (VSBM) for the purpose of information retrieval can be traced back to the seminal paper of Salton et al. [20] who were searching for an appropriate mathematical space to represent documents. Starting from a few basic desiderata, they settled upon a vector in a high dimensional vector space as an appropriate representation of a document. Within this framework, a query is treated like a small (pseudo) document that is also converted to vector form. The documents in the corpus are then ranked according to their distance to the query; closer documents are considered more relevant than ones that are further away. The way was now open to include Boolean operators on the returned results, and thus the first search engines were born. One of the main drawbacks of this system was that it had trouble returning documents that would have been highly relevant if one of the words in the query was replaced by a synonym, and the next advance came from representing concepts latently in a so-called *semantic space* where they are not formally represented or labelled. Semantic spaces are instances of vector spaces, and represent words in a basis created from other words, concepts, documents, or topics. They are generally built from the observation of co-occurrences in large text corpora. In word spaces such as the Hyperspace Analogue to Language (HAL) [21] the basis consists of every word in the vocabulary. Thus, the vector for a given word W is calculated by summing the number of occurrences of word $W(i)$ in a given context window around each occurrence of W and writing that number at the position i in the vector that represents W. This number can be adjusted using the distance (defined in terms of the number of words) or mutual information measures such as Point-Wise Mutual Information, which allows for a weighting of the importance of the word at that position. It is also possible to take word order into account [12,19]. The major evolution with respect to the original proposal of Salton et al., was to derive a more fundamental semantic value through a reduction of the initial word space using mathematical tools such as Singular Value Decomposition [13], Non

Negative Matrix factorization [14], or random projection [18], all of which generate a new basis that is greatly reduced in the number of dimensions. This new basis can under certain conditions be naturally related to topics, objects and concepts [14]. Because of the dimensional reduction, words with similar meaning tend to cluster into single dimensions of the resulting reduced vector space, greatly reducing the problems the old VSBM had with synonyms.

Once a semantic space has been created, we need to rank the results returned by a query using a similarity measure. Several distance measures (such as cosine similarity, Euclidean distance, and the City Block metric [8]) have been applied to semantic analysis, all of which supposedly measure the similarity between words in a given space. The most popular of these in semantic analysis is cosine similarity, which gives the angle between two vectors in a semantic space. We will later explain why this is generally considered a good choice. A number of studies have shown that semantic spaces can be effective at performing tasks that are human like. For example they have shown success at synonymy detection, categorization tasks, information retrieval and query expansion [23]. They have also been shown to perform well at mimicking human word association norms [26]. This success has led a number of researchers to propose semantic spaces as models of human cognition. In this paper we examine important issues related to such a move. Semantic spaces are metric spaces and this poses problems that must be resolved before they can become viable models of human cognition. We shall begin with a discussion of metric spaces and in particular of the properties that a set must satisfy before it can be identified as a metric space. We shall then proceed to a discussion of the way in which human behavior violates these conditions and propose a possible resolution to this problem in later sections.

2.1   Motivating the Angle as a Measure of Similarity

It is notoriously difficult to formally describe the notion of *meaning*. Yet this is precisely what Natural Language Processing aims for. VSBM solve this issue via the so-called *distributional hypothesis*, which claims that *words which occur in similar contexts tend to have similar meanings* [11,10,9]. In VSBM, the entries of the vectors are usually monotone functions of the frequency of co-occurrence. Hence vectors that are "close" occur in similar contexts and, by the distributional hypothesis, ought to have similar meanings. Using the inner product or cosine measure as a representation of similarity then seems like a very plausible suggestion. There are good mathematical reasons as well. If the vectors that correspond to a word are represented by unit vectors, the state space of words becomes the unit sphere. The unit sphere is a simple example of a manifold and geodesics on this manifold are well known to correspond to great circles. On the unit circle, the length of a great circle between two points equals the angle expressed in radians. Indeed, we have that the angle between two points on the sphere is (up to constant scaling) the only unitarily invariant Riemann metric on the sphere [27]. But what precisely are the mathematical criteria for a function to be a bona fide distance function?

2.2   Requirements for a Metric Space

In this section we shall briefly sketch the requirements for a metric space before proceeding in the next section to a discussion of the manner in which semantic data obtained from humans tends to violate these requirements.

Definition 1. *The ordered couple (M, d) with M a non emtpy set and $d : M \times M \to \mathbb{R}$ a function (called the distance or metric), is called a* metric space *if for any $i, j, k \in M$, the following hold:*

1. **Non-negativity:** *the distance between two points must be greater than or equal to zero: $d(i, j) \geq 0$.*
2. **Identity of indiscernibles:** *if the distance between two points is equal to zero then those two points are the same: $d(i, j) = 0 \Leftrightarrow i = j$.*
3. **Symmetry:** *the distance between two points is equal, regardless of which direction it is measured in: $d(i, j) = d(j, i)$.*
4. **The Triangle Inequality:** *for three points in M, the distance from i to k is less than the distance which goes via j: $d(i, j) + d(j, k) \geq d(i, k)$.*

Many authors prefer to list 1 and 2 in a single requirement. In fact, requirement 1 can be derived easily from 2, 3 and 4. It is straightforward to verify that the angle α_{ij} between vectors u_i and u_j:

$$\alpha_{ij} = \cos^{-1} \frac{\langle u_i, u_j \rangle}{|u_i||u_j|}, \tag{1}$$

satisfies all four requirements. The angle between two vectors seems to be in accordance with the distributional hypothesis and satisfies all qualities of a mathematical metric. Moreover, its use has been tested in a wide variety of applications. As such we seem to have a very fundamental and valuable quantity. But the most important question is perhaps how we humans judge semantic similarity. This is a question that belongs to cognitive science so we shall now turn to an examination of similarity in this field, contrasting its results with those of VSBM.

3   Are Semantic Spaces Good Models of Human Cognition?

Vector spaces have been at the heart of many models in cognitive science. One of the more important examples for our purpose, is prototype theory. The basic idea of prototype theory is that some members of a category are more 'typical' than others [17]. For example, a cat is a more (prototypical) member of the category pet, whereas a donkey is clearly more peripheral. This idea is called 'graded categorization' and was formalized by representing concepts as vectors and categories as sets of vectors [15,22]. However, these vectors are not based on co-occurrence, but on subjective numerical scores obtained by questioning human subjects. In this section we shall draw attention to a range of human derived data which violates a number of the properties that must be satisfied by a metric. We shall go through them in the order given in the previous section. The first requirement listed above is non-negativity. This is probably the least problematic of all requirements. Whether or not negative values of similarity occur, is decided by the questionnaire's scale on which human subjects are asked to judge similarity. Humans can quite naturally associate a concept of distance between two words as a measure of their similarity and this distance can be straight-forwardly assumed to be non-negative. However, in this section we shall show that every other requirement of a metric space can be quickly violated by spatial representations of similarity data.

3.1 Homographs and the Non-identity of Indiscernible

The identity of indiscernibles property implies that different words should be separated by some distance. While there are many examples of such a property holding between different words, many languages contain words with multiple meanings, multiple words for the same thing, ambiguous structures, etc. and these properties give us reason to be cautious about its general validity.

For example, we can quickly see that synonyms (different words for the same thing) appear to satisfy the identity of indescernibles property reasonably well; while they lie close together semantically synonyms generally have slightly different connotations. Thus, while 'student' and 'pupil' both mean essentially the same thing, there are slightly different senses to these two words, and hence they tend to appear close together, but with some distance separating them in most semantic spaces. In contrast, homographs create much more serious problems for attempts to generate a metric space. Homographs are words that have the same spelling and pronunciation but different meanings. For example, 'bat' is a homograph, as it has at least two senses: (1) as a small furry flying mammal; and (2) as a sporting implement.

Homographs pose a problem for the *if and only if* criterion in property 2. If we generate a set that represents each word in English, then 'bat' should appear only once in it ($i = j$); however, semantic spaces tend to correctly reveal the different meanings behind this word by using a mixture of the representation of both words. Thus, property 2 seems to pose a challenge for semantic space approaches, as discernible words (such as 'bat' for sports and 'bat' the animal) are represented at exactly the same point in the space. We believe a finer resolution of homographs in semantic space is possible by examining the set of documents that contain the words. First a search in, for example, Wordnet will reveal if a word has several meanings and if so, how many. Say a word has n possible meanings. Then we ought to divide the set of all the words that substantially co-occur with the query word, into n sets of words such that each set shows a degree of cohesion in the words that co-occur with it. This may be implemented by an appropriate algorithm that reduces to n the dimension of the matrix that has as its rows the words that co-occur and as columns the documents in which they occur. Interestingly, a very similar situation occurs in quantum mechanics in the case of degenerate energy levels. An energy level of a quantum system is called degenerate if different states correspond to the same energy level. If we think of the energy level of the system as 'the name' of the state that corresponds to that energy level, we have an analogy with homographs. Application of a well chosen perturbation to the Hamiltonian of the system allows us to separate the energy levels, so each energy level corresponds in a unique way to an energy level. We say that the perturbing field is 'lifting the degeneracy' and splits the energy level into finer energetic detail. If we see a separation of the two meanings of a single word in the semantic vector space, it seems we have provided enough context in the semantic space to lift the degeneracy of meanings corresponding to a single word. In an actual task of information retrieval, it is very valuable to be able to identify which meaning is more probable for a given word in a given context. For this we would have to judge to which of two statistical clusters a given vector (word) in a given context belongs. Language is extremely flexible and is perfectly able to shift perspective as we include more context, thereby changing the meaning. Take as an example, the word 'hits'. Without additional context, its meaning is degenerate; it could mean many things. We are then given a piece of context: 'Michael hits Billy'. Most probably 'hits' denotes a physical act of violence. We are then given an additional piece of context: 'Michael Jackson hits Billy Jean'. The meaning of 'hits' is now more likely to signify a musical hit. We are given a last piece of context: 'Michael Jackson number of Google hits for Billy Jean', the word 'hits' denotes the webpages Google relates to a query. In the example above every new level of context only adds words to the previous context; the previous context isn't changed in form, only in meaning. We feel the nature of language is simply too rich to allow for a strict separation, but VSBM do seem capable of at least statistically approaching the problem of homographs.

3.2   Human Similarity Judgements Are Not Symmetric

It was shown by Tversky that human similarity judgements are asymmetric, and so directly violate the symmetry requirement of metric spaces (i.e. $d(a, b) \neq d(b, a)$) [25]. A classic example was first provided by Rosch in her theory of prototypes [16], which shows that humans have a tendency to declare similarity with respect to an archetype. For example, when asked to give an example of the concept furniture, humans will much more frequently nominate a "chair" than a "stool", and this archetypical concept ("chair") is the one that similarity judgements are preferentially, and asymmetrically, assigned by. Thus, the similarity of stool to chair is usually deemed to be greater than that of chair to stool, the similarity of North Korea to China is judged greater than the similarity of China to North Korea [24,25], and pink is deemed more similar to red, than red is to pink. This seems to be a genuine linguistic phenomenon that one would eventually like to model. Of course, these experiments are designed to test for asymmetry; experiments that do not show asymmetry are equally easy to design. Suppose we produce a deck of cards with on each card nothing but the two words "red" and "pink". However, on half of the cards the word "red" is printed above the word "pink", on the other half, "pink" is printed above "red". Each test subject is given one card and asked to quantify the similarity of the two concepts printed on the card. The result will obviously be symmetrical, because there was no distinguished order of words on the deck of cards. For our present purpose, we will assume symmetrical data.

3.3 Human Similarity Judgements Violate the Triangle Inequality

Finally, human similarity judgements do not appear to satisfy the triangle inequality, a result shown by Tversky & Gati [24]. Indeed, the contrast between human similarity judgements and distance notions in geometric models of cognition led them to conclude that ([24], p 153):

> *some basic properties of the geometric model (e.g., translation invariance, segmental additivity, and the triangle inequality), which enhance the interpretability and the appeal of spatial representations, cannot always be accepted as valid principles of psychological similarity.*

even before Semantic Space approaches to the mental lexicon were invented.

If Tversky & Gati are correct then their criticism poses some very serious problems for both semantic space, and hence their associated quantum inspired, models of the human mental lexicon. To put things in perspective, semantic spaces were developed and successfully put to use in spite of this problem, so perhaps we need not worry too much. However, we would like to be able to model subjective similarity, as it seems to be an important component of natural language processing. What makes the triangle inequality problem more severe than the three previous requirements we discussed, it that we cannot make it go away by devising another experiment, at least not straightforwardly. If we want symmetric or non-negative data we can always make sure that the experiment will give us only positive values. For non-negativity we need only to constrain the range of the possible answers; for the symmetry condition, we need only to make sure every couple's similarity is symmetric. Indeed, if $d(a, b) = d(b, a)$ and $d(b, c) = d(c, b)$, then obviously $d(a, c) = d(c, a)$. Can we design an experiment in such a way that it always satisfies the triangle inequality? We could give concepts in triples to subjects and ask them to draw a triangle with the three words on the vertices of the triangle and express the relative similarities by the relative lengths of the sides of the triangle. The triangle inequality would be trivially satisfied for this triple. However, if we have several triples that satisfy the triangle inequality, then there is no guarantee whatsoever, that from these triples we cannot pick words to form new triples that will violate the triangle inequality. Another proposal would be to abandon metric spaces, or geometric models for the representation of cognitive entities such as concepts and sentences. If we take into consideration the huge success this class of models has enjoyed then this seems like a rather radical step to take. An alternative answer to Tversky & Gati might be found through an adoption of the notion of context, and in what follows we shall start to develop an approach within a metric space that can recover the non-metric behavior of human similarity judgements.

4   The Point of View Model

In vector space based accounts of cognition (such as quantum theory inspired approaches [1,2,3,5,6,7]) concepts are very often represented by unit vectors in a Hilbert space. Take three unit vectors u_1, u_2 and u_3 that represent three concepts. Calling θ_{ij} the angle between u_i and u_j, we find that

$$\cos\theta_{ij} = \langle u_i, u_j \rangle. \tag{2}$$

Because Hilbert space is a metric space, this has consequences for the possible range of values the angles between the vectors can assume:

$$|\theta_{ij} - \theta_{jk}| \leq \theta_{ik} \leq |\theta_{ij} + \theta_{jk}|. \tag{3}$$

The point of view model assumes that each time a subject is asked to quantify the similarity between two concepts they must take a stance, or a *point of view*, from which to judge their similarity. On an absolute scale we may argue that all concepts are very similar (they are, after all, just concepts), or we may argue no two concepts are alike. But if we are asked what the similarity is between *Moon* and *Ball*, we will not easily judge their similarity on an absolute scale. We rather inadvertently look for a proper context to judge their similarity. If our perspective is *"Shape"* then we will think of *Moon* and *Ball* as being somewhat similar. If the perspective would have been *"Play"*, the two concepts would be judged rather dissimilar. So it is the two words, together with the state of the subject, that determine the point of view from where similarity will be judged. We model a point of view by assuming that for each pair of vectors u_i and u_j and a given subject S that is asked to judge their similarity, there is a *point of view vector* (POVV) u_{ij}^S. The cosine of the angle this observer sees between u_i and u_j, is:

$$\begin{aligned}
\cos\alpha_{ij} &= \frac{\langle u_i - u_{ij}^S, u_j - u_{ijj}^S \rangle}{|\langle u_i - u_{ij}^S \rangle||\langle u_j - u_{ij}^S \rangle|} \\
&= \frac{\cos\theta_{ij} - \langle u_{ij}^S, u_j \rangle - \langle u_i, u_{ij}^S \rangle + |u_{ij}^S|}{|\langle u_i - u_{ij}^S \rangle||\langle u_j - u_{ij}^S \rangle|}.
\end{aligned} \tag{4}$$

In psychological experiments, the similarity is an average over many trials. The expected similarity is then derived as a measure over the POVV's. In what follows, we may assume that u_{ij}^S is already an averaged point of view in the sense that α_{ij} coincides with the average subjective similarity. To determine which regions for u_{ij}^S lead to increased values of θ_{ij} and which lead to decreased values, we first look at the set of u_{ij}^S that leaves α_{ij} invariant.

Lemma 1. *Let $0, u_i$ and u_j be three non-collinear vectors and let C_{ij} be the circle that contains $0, u_i$ and u_j. Then for any $u_{ij}^S \in C_{ij}$ with $u_{ij}^S \neq u_i$ and $u_{ij}^S \neq u_j$ we have $\cos\alpha_{ij} = \cos\theta_{ij}$.*

Proof. The span u_i and u_j defines a two dimensional linear subspace containing the null vector. Let C_{ij} be the unique circle within this linear subspace that contains 0, u_i and u_j. By the inscribed angle theorem –which states that an angle inscribed in a circle is half of the central angle that subtends the same arc on the circle–, the angle θ_{ij} inscribed in this circle does not change as its apex u_{ij}^S is moved to different positions on C_{ij}, hence $\alpha_{ij} = \theta_{ij}$.   □

Now that we have fixed the region for which the observed similarity remains invariant, we will look at the interval of values the similarity can take.

Lemma 2. *Given two concepts c_i and c_j, represented by two vectors u_i and u_j, there exists a point of view vector such that the observed angle α_{ij} can take values in the interval $[\frac{1}{2}\arccos\langle u_i, u_j\rangle, \pi]$.*

To see this is indeed the case, call D_{ij} the open disk that is the interior of C_{ij}. It is easy to see a POVV inside D_{ij} yields an observed angle α_{ij} that is greater than θ_{ij}. The disk D_{ij} is an open convex set, so any open convex combination of 0, u_i and u_j is an element of D_{ij}. The maximal angle is reached for $u_{ij}^S = \frac{1}{2}(u_i + u_j)$, which clearly lies inside D_{ij}. The observed angle in this case is

$$\alpha_{ij} = \cos^{-1} \frac{\langle u_i - \frac{1}{2}(u_i + u_j), u_j - \frac{1}{2}(u_i + u_j)\rangle}{|\langle u_i - \frac{1}{2}(u_i + u_j)\rangle||\langle u_j - \frac{1}{2}(u_i + u_j)\rangle|} \tag{5}$$

$$= \cos^{-1}(-1) = \pi. \tag{6}$$

So it is always possible to pick a POVV in D_{ij} that yields the minimal similarity. (This result makes sense geometrically: if your point of view is in the middle of the two concepts, then, to you, they couldn't be further apart from each other.) For an intermediate situation, there are many possibilities. A particularly nice choice is to consider the $d-$parameter POVV that lies precisely between u_i and u_j and has length d : $u_{ij}^S(d) = \frac{d}{|u_i+u_j|}(u_i + u_j)$. If we consider the triangle which has as vertices $u_{ij}^S(d), o$ and u_i, the sine rule immediately tells us that $\sin(\pi - \alpha_{ij}/2)/1 = \sin((\alpha_{ij} - \theta_{ij})/2)/d$, hence the relation between α_{ij}, θ_{ij} and d is given by

$$d = \frac{\sin((\alpha_{ij} - \theta_{ij})/2)}{\sin(\alpha_{ij}/2)}. \tag{7}$$

We can get minimal similarity and intermediate values. It turns out the POVV constrains the maximum similarities (minimal angle). To see this, take

$$u_{ij}^S = -\frac{u_i + u_j}{|u_i + u_j|}. \tag{8}$$

This unit vector points in the direction opposite of $\frac{1}{2}(u_i + u_j)$. We have again (using the inscribed angle theorem) that the observed angle α_{ij} is exactly $\theta_{ij}/2$. This is obviously the minimal value for α_{ij} that the point of view model can attain; it is reached if u_{ij}^S lies on the great arc of the unit circle between u_i and u_j. So we have demonstrated that there exist POVV such that the observed angle α_{ij} can be as high as $\theta_{ij}/2 = \frac{1}{2}\arccos\langle u_i, u_j\rangle$.

4.1   The Evocation Data Set

Let us provide a brief illustration of the model using data from the Evocation data set [4], collected by crowd sourcing using Amazon Mechanical Turk (which allows for the quick collection of large amounts of data). The data was cleaned to the highest level of correlation with a smaller data set collected under controlled conditions. Users were asked how much a sense of a word brings the sense of another to mind (on a scale of 0 to 100), using the words as well as a definition for disambiguation. The data for a pair of words are usually not symmetric, however for the purposes of this paper we have averaged the two similarities so that the resulting data is symmetric. In essence then, this data set contains human judgements of symmetrized semantic relatedness between pairs of words. For example, *'key'* and *'car'* were judged at 73% of similarity, *'car'* and *'light'* at 79,4% of semantic similarity, while *'key'* and *'light'* only at 14.3%. Other examples of triples that violate the triangle inequality from this data set include:

1. *night /day*: 86.3%, *day/year*: 62.8%, *night/ year*: 11.6%;
2. *school/university*: 83.7%, *university/court*: 73.2%, *school/court*: 7.6%;
3. *food/oil*: 81.5%, *oil/gold*: 62.8%, *food/gold*: 2.7%.

Let us take the first example and label three vectors with an index that refers to the concepts: u_n is the vector that corresponds to *night*, and likewise we denote u_d for the concept *day* and u_y for the concept *year*. We first convert the given similarities to angles using $\cos\theta_{ij} = \langle u_i, u_j \rangle$. Then $\theta_{nd} = 0.53; \theta_{dy} = 0.89$ and $\theta_{ny} = 1.45$. Clearly this triple violates the triangle inequality, e.g. $|\theta_{ny} - \theta_{dy}| = |1.45 - 0.89| = 0.56 \geq \theta_{nd} = 0.53$. Because the triangle inequality is violated, there do not exist three vectors with the prescribed angles. However, from the $d-$parametrized POVV for θ_{ny}, $u_{ij}^S(d) = \frac{d}{|u_n + u_y|}(u_n + u_y)$, we obtain: $|u_n + u_y| = 2\cos(\theta_{ny}/2) \approx 1.5$. The value of θ_{ny} was 1.45; if it would have been 1.42, no violation would have occurred. Hence we choose $d = \sin((1.42 - 1.45)/2)/\sin(1.42/2) = -.023$. So the POVV $u_{ij}^S(d) = \frac{-1}{60}(u_n + u_y)$ restores the triangle inequality for this triple. It is easy to see we could also have taken a triple of vectors that respect the inequality (e.g., the "restored" vectors above) and, when one of the angles is viewed upon from a suitably chosen POVV (e.g., the opposite vector of $u_{ij}^S(d)$ in the example above), the resulting angles will violate the inequality.

5   Concluding Remarks

The question we addressed in this paper is whether it is possible for a semantic space to be a metric space and at the same time be able to capture the non-metric behavior of human similarity judgements. Another strongly related and perhaps even more interesting question is whether it is possible to derive a vector space using subjective similarity instead of co-occurrence. We presented a model that gives an affirmative answer, in principle. Although the model we offered here was derived in an essentially *ad hoc* way, the model is falsifiable and we feel the case for this model could be made stronger if it can be shown a POVV can be derived from the semantic space itself. In order to sketch out a viable avenue for further work, we shall refer to one of our above examples. It is not peculiar that *day* and *year* are considered close, as they are both important measures of time. Neither is it strange that *day* and *night* are judged to be close, as they are in a certain sense opposite to one another. Note that someone who is being asked how close *day* and *night* are, will think of *day* in the sense of daytime, which is not the same meaning the word has when we compare *day* and *year*. The last couple in our triple is then *night* and *year*, which are not so obviously connected, hence the lower similarity rating. We see that when we are asked to weigh the words for similarity, we unconsciously look for a minimal context that contains the two concepts, and depending on the words, this will be a different context. This is what the POVV model attempts to capture. However, for the POVV model to be convincing, we need to show there is a connection between the POVV and the concepts we are dealing with. In particular, the vectors that correspond to the words and their semantically associated vectors should determine the POVV. In a sense, the POVV is a "centre of gravity of meaning": if all concepts contribute to the centre of gravity, then the POVV will approximately be the zero of the vector space and the triangle inequality will hold; if not, deviations will arise. An important observation is that the model as it is right now, does not specify a unique POVV, so how will we know an eventual linkage between pairs of words and POVV's is viable? A valid confirmation would require a statistically significant test that uses only a semantic network and no human similarity measures, and which can predict human violations of the triangle inequality for triples of words. Whether this avenue will prove fruitful is left for future research.

Acknowledgements. This project was supported by the Marie Curie International Research Staff Exchange Scheme: Project 247590, "QONTEXT - Quantum Contextual Information Access and Retrieval". KK is supported by the Australian Research Council Discovery grant DP1094974.

References

1. Aerts, D., Gabora, L.: A theory of concepts and their combinations I: the structure of the sets of contexts and properties. Kybernetes 34, 151–175 (2005)
2. Aerts, D., Gabora, L.: A theory of concepts and their combinations II: A Hilbert space representation. Kybernetes 34, 192–221 (2005)
3. Aerts, D.: Quantum structure in cognition. Journal of Mathematical Psychology 53, 314–348 (2009)
4. Boyd-Graber, J., Fellbaum, C., Osherson, D., Schapire, R.: Adding dense, weighted connections to wordnet. In: Proceedings of the Third International WordNet Conference (2006)
5. Bruza, P.D., Kitto, K., Ramm, B., Sitbon, L.: The non-decomposability of concept combinations (2011) (under review)
6. Bruza, P.D., Kitto, K., Ramm, B., Sitbon, L., Blomberg, S., Song, D.: Quantum-like non-separability of concept combinations, emergent associates and abduction. Logic Journal of the IGPL (2010) (in press)

7. Bruza, P., Kitto, K., Nelson, D., McEvoy, C.: Is there something quantum-like about the human mental lexicon? Journal of Mathematical Psychology 53, 362–377 (2009)
8. Bullinaria, J.A., Levy, J.P.: Extracting semantic representations from word co-occurrence statistics: a computational study. Behavior Research Methods 39(3), 510–526 (2007)
9. Deerwester, S., Dumais, S., Furnas, G., Landauer, T., Harshman, R.: Indexing by latent semantic analysis. Journal of the American Society for Information Science 41(16), 391–407 (1990)
10. Firth, J.R.: Papers in Linguistics, pp. 1934–1951. Oxford University Press, London (1957)
11. Harris, Z.: Distributional structure. Word 10(23), 146–162 (1954)
12. Jones, M.N., Mewhort, D.J.K.: Representing word meaning and order information in a composite holographic lexicon. Psychological Review 114(1), 1–37 (2007)
13. Landauer, T., Dumais, S.T.: A solution to plato's problem: the latent semantic analysis theory of acquisition, induction and representation of knowledge. Psychological Review 104(2), 211–240 (1997)
14. Lee, D.D., Seung, H.S.: Learning the parts of objects by non-negative matrix factorization. Nature 401(6755), 788–791 (1999)
15. Nosofsky, R.M.: Attention, similarity, and the identification-categorization relationship. Journal of Experimental Psychology: General 115(1), 39–57 (1986)
16. Rosch, E.: Cognitive Representation of Semantic Categories. Journal of Experimental Psychology 104, 192–233
17. Rosch, E., Lloyd, B.B. (eds.): Cognition and categorization. Erlbaum, Hillsdale (1978)
18. Sahlgren, M.: An introduction to random indexing. In: Proceedings of Methods and Applications of Semantic Indexing Workshop at the 7th International Conference on Terminology and Knowledge Engineering, Copenhagen, Denmark (2005)
19. Sahlgren, M., Holst, A., Kanerva, P.: Permutations as a means to encode order in word space. In: Proceedings of the 30th Annual Meeting of the Cognitive Science Society (CogSci 2008), Washington, D.C., USA, July 23-26 (2008)
20. Salton, G., Wong, A., Yang, C.S.: A vector space model for automatic indexing. Communications of the ACM 18(11), 613–620 (1975)
21. Schütze, H.: Automatic word sense discrimination. Computational Linguistics 24(1), 97–123 (1998)
22. Smith, E.E., Osherson, D.N., Rips, L.J., Keane, M.: Combining prototypes: A selective modification model. Cognitive Science 12(4), 485–527 (1988)
23. Turney, P.T., Pantel, P.: From frequency to meaning: Vector space models of semantics. Journal of Artificial Intelligence Research 37, 141–188 (2010)
24. Tversky, A., Gati, I.: Similarity, separability, and the triangle inequality. Psychological Review 89(2), 123–154 (1982)
25. Tversky, A.: Features of similarity. Psychological Review 84(4), 327–352 (1977)
26. Veksler, V.D., Govostes, R.Z., Gray, W.D.: Defining the dimensions of the human semantic space. In: Sloutsky, V., Love, B., McRae, K. (eds.) 30th Annual Meeting of the Cognitive Science Society, pp. 1282–1287. Cognitive Science Society (2008)
27. Wooters, W.K.: The Acquisition of Information from Quantum Measurements. Ph.D. thesis, University of Texas at Austin (1980)

Toward a Formal Model of the Shifting Relationship between Concepts and Contexts during Associative Thought

Tomas Veloz[1], Liane Gabora[1], Mark Eyjolfson[1], and Diederik Aerts[2]

[1] University of British Columbia, Department of Psychology, Okanagan campus, 3333 University Way Kelowna BC, V1V 1V7, Canada
{tomas.veloz,liane.gabora}@ubc.ca, mark_eyjolfson@hotmail.com
[2] Center Leo Apostel for Interdisciplinary Studies, Vrije Universiteit Brussel, Belgium
diraerts@vub.ac.be

Abstract. The quantum inspired State Context Property (SCOP) theory of concepts is unique amongst theories of concepts in offering a means of incorporating that for each concept in each different context there are an unlimited number of exemplars, or states, of varying degrees of typicality. Working with data from a study in which participants were asked to rate the typicality of exemplars of a concept for different contexts, and introducing a state-transition threshold, we built a SCOP model of how states of a concept arise differently in associative versus analytic (or divergent and convergent) modes of thought. Introducing measures of expected typicality for both states and contexts, we show that by varying the threshold, the expected typicality of different contexts changes, and seemingly atypical states can become typical. The formalism provides a pivotal step toward a formal explanation of creative thought processes.

Keywords: Associative thought, concepts, context dependence, contextual focus, creativity, divergent thinking, dual processing, SCOP.

1   Introduction

This paper unites two well-established psychological phenomena using a quantum-inspired mathematical theory of concepts, the State-COntext-Property (SCOP) theory of concepts. The first phenomenon is that the meaning of concepts shifts, sometimes radically, depending on the context in which they appear [19, 13, 9]. It is this phenomenon that SCOP was developed to account for [3, 4, 5]. Here we use SCOP to model a different though related psychological phenomenon. This second psychological phenomenon was hinted at in the writings of a number of the pioneers of psychology, including Freud [17], Piaget [10], and William James [20]. They and others have suggested that all humans possess two distinct ways of thinking. The first, sometimes referred to as divergent or *associative thought*, is thought to be automatic, intuitive, diffuse, unconstrained, and conducive to unearthing remote or subtle associations between items that share features, or that are *correlated* but not necessarily *causally* related. This may yield a promising idea or solution though perhaps in a vague, unpolished form. There is evidence that associative thinking involves controlled access to, and integration of, affect-laden material, or what Freud referred to as "primary process" content [17,18]. Associative thought is contrasted with a more controlled, logical,

D. Song et al. (Eds.): QI 2011, LNCS 7052, pp. 25–34, 2011.

rule-based, convergent, or *analytic* mode of thought that is conducive to analyzing relationships of cause and effect between items already believed to be related. Analytic thought is believed to be related to what Freud termed "secondary process" material.

A growing body of experimental and theoretical evidence for these two modes of thought, associative and analytic, led to hypothesis that thought varies along a continuum between these two extremes depending on the situation we are in [7, 15, 17, 3, 11, 13, 14, 20]. The capacity to shift between the two modes is sometimes referred to as *contextual focus*, since a change from one mode of thought to the other is is brought about by the context, through the focusing or defocusing of attention [11, 12]. Contextual focus is closely related to the dual-process theory of human cognition, the idea that human thought employs both implicit and explicit ways of learning and processing information [16, 8]. It is not just the existence of two modes of thought but the cognitive consequences of shifting between them, that we use SCOP to model in this paper.

2   The SCOP Theory of Concepts

The SCOP formalism is an operational approach in the foundations of quantum mechanics in which a physical system is determined by the mathematical structure of its set of states, set of properties, the possible (measurement) contexts which can be applied to this entity, and the relations between these sets. The SCOP formalism is part of a longstanding effort to develop an operational approach to quantum mechanics known as the Geneva-Brussels approach [1]. If a suitable set of quantum axioms is satisfied by the set of properties, one recovers via the Piron-Solèr representation theorem the standard description of quantum mechanics in Hilbert space [1]. The SCOP formalism permits one to describe not only physical entities, but also potential entities [2], which means that SCOP aims at a very general description of how the interaction between context and the state of an entity plays a fundamental role in its evolution. In this work we make use of the SCOP formalism to model concepts, continuing the research reported in [4, 5, 3, 6].

Formally a conceptual SCOP entity consists of three sets Σ, $\mathcal{M}$, and $\mathcal{L}$: the set of states, the set of contexts and the set of properties, and two additional functions μ and ν. The function μ is a probability function that describes how state p under the influence of context e changes to state q. Mathematically, this means that μ is a function from the set $\Sigma \times \mathcal{M} \times \Sigma$ to the interval $[0,1]$, where $\mu(q,e,p)$ is the probability that state p under the influence of context e changes to state q. We write

$$\mu : \Sigma \times \mathcal{M} \times \Sigma \to [0,1]$$
$$(q,e,p) \mapsto \mu(q,e,p) \tag{1}$$

The function ν describes the weight, which is the renormalization of the applicability, of a certain property given a specific state. This means that ν is a function from the set $\Sigma \times \mathcal{L}$ to the interval $[0,1]$, where $\nu(p,a)$ is the weight of property a for the concept in state p. We write

$$\nu : \Sigma \times \mathcal{L} \to [0,1]$$
$$(p,a) \mapsto \nu(p,a) \tag{2}$$

Thus the SCOP is defined by the five elements $(\Sigma, \mathcal{M}, \mathcal{L}, \mu, \nu)$. States of a concept are denoted by means of the letters $p, q, r, \ldots$ or $p_1, p_2, \ldots$, and contexts by means of the letters $e, f, g, \ldots$ or $e_1, e_2, \ldots$. When a concept is not influenced by any context, we say is in its *ground state*, and we denote the ground state by $\hat{p}$. The unit context, denoted 1, is the absence of a specific context. Hence context 1 leaves the ground state $\hat{p}$ unchanged. Exemplars of a concept are states of this concept in the SCOP formalism.

Note that in SCOP, concepts exist in what we refer to as a state of potentiality until they are evoked or actualized by some context. To avoid misunderstanding we mention that $\mu(p, e, q)$ is not a conditional probability of transitioning from state p to q *given that* the context is e. Contexts in SCOP are not just conditions, but active elements that *alter* the state of the concept, analogous to the *observer phenomenon* of quantum physics, where measurements affect the state of the observed entity. Indeed, a SCOP concept can be represented in a complex Hilbert space $\mathcal{H}$. Each state p is modelled as a unitary vector (pure state) $|p\rangle \in \mathcal{H}$, or a trace-one density operator (density state) ρ_p. A context e is generally represented by a linear operator of the Hilbert space $\mathcal{H}$, that provokes a probabilistic collapse by a set of orthogonal projections $\{P_i^e\}$. A property a is always represented by an orthogonal projector P_a in $\mathcal{H}$ respectively. The contextual influence of a context on a concept is modelled by the application of the context operator on the concept's state. A more detailed explanation can be found in [4, 5].

3   The Study

Our application of SCOP made use of data obtained in a psychological study of the effect of context on the typicality of exemplars of a concept. We now describe the study.

3.1   Participants and Method

Ninety-eight University of British Columbia undergraduates who were taking a first-year psychology course participated in the experiment. They received credit for their participation.

The study was carried out in a classroom setting. The participants were given questionnaires that listed eight exemplars (states) of the concept HAT. The exemplars are: state p_1: 'Cowboy hat', state p_2: 'Baseball cap', state p_3: 'Helmet', state p_4: 'Top hat', state p_5: 'Coonskincap', state p_6: 'Toque', state p_7: 'Pylon', and state p_8: 'Medicine Hat'. They were also given five different contexts. The contexts are: the default or unit context e_1: *The hat*, context e_2: *Worn to be funny*, context e_3: *Worn for protection*, context e_4: *Worn in the south*, and context e_5: *Not worn by a person*.

The participants were asked to rate the typicality of each exemplar on a 7-point Likert scale, where 0 points represents "not at all typical" and 7 points represents "extremely typical". Note that all the contexts except e_1 make reference to the verb "wear", which is relevant to the concept HAT. The context e_1 is included to measure the typicality of the concept in a context that simulates the *pure* meaning of a HAT, i.e. having no contextual influence, hence what in SCOP is meant by "the unit context".

3.2   Results

A summary of the participants' ratings of the typicality of each exemplar of the concept HAT for each context is presented in Table 1. The contexts are shown across the top, and

Table 1. Summary of the participants' ratings of the typicality of the different exemplars of the concept HAT for different contexts. See text for detailed explanation.

Exp. Data	e_1	e_2	e_3	e_4	e_5
p_1 Cowboy hat	(5.44;0.18)	(3.57;0.14)	(3.06;0.13)	(6.24;0.28)	(0.69;0.05)
p_2 Baseball cap	(6.32;0.21)	(1.67;0.06)	(3.16;0.13)	(4.83;0.21)	(0.64;0.04)
p_3 Helmet	(3.45;0.11)	(2.19;0.08)	(6.85;0.28)	(2.85;0.13)	(0.86;0.06)
p_4 Top hat	(5.12;0.17)	(4.52;0.17)	(2.00;0.08)	(2.81;0.12)	(0.92;0.06)
p_5 Coonskincap	(3.55;0.11)	(5.10;0.19)	(2.57;0.10)	(2.70;0.12)	(1.38;0.1)
p_6 Toque	(4.96;0.16)	(2.31;0.09)	(4.11;0.17)	(1.52;0.07)	(0.77;0.05)
p_7 Pylon	(0.56;0.02)	(5.46;0.21)	(1.36;0.05)	(0.68;0.03)	(3.95;0.29)
p_8 Medicine Hat	(0.86;0.02)	(1.14;0.04)	(0.67;0.03)	(0.56;0.02)	(4.25;0.31)
$N(e)$	30.30	25.98	23.80	22.22	13.51

exemplars are given in the left-most column. For each state and context in the table there is a pair of numbers $(a;b)$. a represents the averaged sum of the Likert points across all participants (average typicality). b is the context dependent state-transition probability. The bottom row gives the normalization constant of each transition probability function. Grey boxes have transition probability below the threshold $\alpha = 0.16$.

4   Analysis of Experimental Data and Application to the Model

In this section we use SCOP to analyze the data collected in the experiment, and apply it to the development of a tentative formal model of how concepts are used differently in analytic and associative thought.

4.1   Assumptions and Goals

We model the concept HAT by the SCOP $(\Sigma, \mathcal{M}, \mathcal{L}, \mu, \nu)$ where $\Sigma = \{p_1, \ldots, p_8\}$ and $\mathcal{M} = \{e_1, \ldots, e_5\}$ are the sets of exemplars and contexts considered in the experiment (see table 1). We did not consider properties of the concept HAT, and hence $\mathcal{L}$ and ν are not specified. This is a small and idealized SCOP model, since only one experiment with a fairly limited number of states and contexts is considered, but it turned out to be sufficient to carry out the qualitative analysis we now present. Moreover, it will be clear that the approach can be extended in a straightforward way to the construction of more extended SCOP models that include the applicabilities of properties. Note also that the Hilbert space model of this SCOP can be constructed following the procedure explained in [5].

Recall how the participants estimated the typicality of a particular exemplar p_i, $i \in \{1, \ldots, 8\}$ under a specified context e_j, $j \in \{1, \ldots, 5\}$ by rating this typicality from 0 to 7 on a Likert scale. Since these ratings play a key role in the analysis, we introduce the Likert function L:

$$L : \Sigma \times \mathcal{M} \rightarrow [0,7] \tag{3}$$

$$(p,e) \mapsto L(p,e) \tag{4}$$

where $L(p,e)$ is the Likert score averaged over all subjects given to state p under context e.

We also introduce the total Likert function N which gives the total Likert score for a given context:

$$N : \mathcal{M} \to [0,56]$$
$$e \mapsto N(e) = \sum_{p \in \Sigma} L(p,e), \tag{5}$$

The Likert score $L(p,e)$ is not directly connected to the transition probability $\mu(p,e,\hat{p})$ from the ground state of a concept to the state p under context e. However, the renormalized value of $L(p,e)$ to the interval $[0,1]$

provides a reasonable estimate of the transition probability $\mu(p,e,\hat{p})$. Hence we introduce the hypothesis that the renormalized Likert scores correspond to the transition probabilities from the ground state, or

$$\mu(p,e,\hat{p}) = \frac{L(p,e)}{N(e)} \tag{6}$$

This is an idealization since the transition probabilities are independent although correlated to this renormalized Likert scores. In future work we plan experiments to directly measure the transition probabilities.

Let us pause briefly to explain why these functions have been introduced. If we consider the unit context, it would be natural to link the typicality to just the Likert number. For example, for the unit context, exemplar p_1: 'Cowboy hat' is more typical than p_6: 'Toque' because $L(p_6,e_1) < L(p_1,e_1)$ (see table 1). If one examines more than one context, however, such a conclusion cannot easily be drawn. For example, consider the exemplar p_7: 'Pylon', under both the context e_2:*Worn to be funny* and context e_5: *Not worn by a person*, we have that $L(p_7,e_5) < L(p_7,e_2)$, but p_7 is more typical under context e_5 than under e_2. This is because $N(e_5) < N(e_2)$, i.e. the number of Likert points given in total for context e_2 is much higher than the number of Likert points given in total for the context e_5. This is primarily due to the fact that Likert points have been attributed by participant per context.

Note that $\frac{N(e)}{8}$ is the average typicality of exemplars under context e, and the average transition probability (renormalized typicality) is $\mu^* = \frac{1}{8}$ for all the contexts. We want to identify the internal structure of state transitions of a concept making use of the typicality data. Therefore we define a transition probability threshold $\alpha \in [0,1]$. We say that $p \in \Sigma$ is *improbable* for context $e \in \mathcal{M}$ if and only if $\mu(p,e,\hat{p}) < \alpha$, meaning that it is improbable that a transition will happen under this context to states with transition probability lower than the threshold. By means of this transition threshold we can also express the idea that for a given concept, there are only a limited number of possible transitions from the ground state to other states. We express this mathematically by introducing a new collection of transition probabilities, such that for this new collection the transition probability is equal to zero when it is below this threshold, thereby prohibiting transitions from a specific context to states that we called improbable for this context for the original collection of transition probabilities we started with. Since the sum of all transition probabilities over all possible states that can be transitioned to needs to be equal to 1 for any set of transition probabilities corresponding to an experimental situation, next to equaling to zero the transition probabilities below the threshold, we need to renormalize the remaining transition probabilities. Hence, if we denote μ_α the new collection of transition probabilities, we have

$$\mu_\alpha(p,e,\hat{p}) = 0 \quad \text{if} \quad \mu(p,e,\hat{p}) \le \alpha, \text{else} \tag{7}$$

$$= \frac{\mu(p,e,\hat{p})}{\sum_{p\in\Sigma, \alpha<\mu(p,e,\hat{p})} \mu(p,e,\hat{p})} \tag{8}$$

Thus, after imposing a threshold, a concept becomes a more constrained structure. At first glance this may appear to be an artificial bias in our analysis. However, we do not introduce the threshold to arbitrarily eliminate some exemplars, but to study the evolution of this *biased structure* as the threshold changes. This leads to the next step, which is to model what happens to the exemplars and contexts when there is a shift between associative and analytic thought modes of thought.

For each exemplar p and context e such that $\mu(p,e,\hat{p}) > \alpha$ we have that $\mu_\alpha(p,e,\hat{p}) > \mu(p,e,\hat{p})$. The new collection of transition probabilities induced by α corresponds to the fact that in an associative mode we gain access to remote meanings while in an analytic mode of thought we lose them. Hence, the transition probability to an unusual exemplar p, which is zero for a high setting of transition probabilities (and thus considered a *strange exemplar* for the concept within this setting) could rise above zero for the new α-induced setting of transition probabilities. This occurs when the strange exemplar p is typical *compared to* other exemplars under context e, i.e. $\mu(p,e,\hat{p})$ is high enough. Thus, one shifts to a more associative mode of thought by decreasing the threshold, thereby enabling unusual exemplars to come into play. We propose that this is the mechanism that underlies contextual focus [3, 11, 12].

5   Analysis of the States and Contexts

5.1   Expected Context Typicality

Since the SCOP model is a probabilistic model, the typicalities estimated by the participants in the experiment by numbers on the Likert scale are not the expected typicalities, because the transition probabilities must also be taken into account. This expresses the potentiality (and corresponding probability), which is fundamental to the SCOP approach. Indeed, it makes only sense to speak of the "potential typicality" of a certain exemplar, and this potentiality is expressed by the value of the transition probability to this exemplar, which means that this "potential typicality" is the "expected typicality" which equals to the product of the Likert value with the transition probability, i.e. $L(p,e) \cdot \mu_\alpha(p,e,\hat{p})$. This provides now also a means of introducing a genuine measure of context typicality, using the state transition probability model, and the *mode of thought* determined by the threshold α. For a given context e and a given threshold α the "expected typicality $T(e,\alpha)$ of this context e" is given by

$$T(e,\alpha) = \sum_{p\in\Sigma} L(p,e) \cdot \mu_\alpha(p,e,\hat{p}) \tag{9}$$

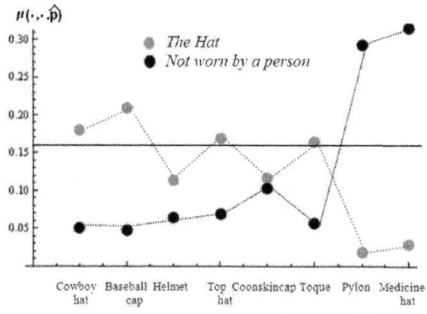

Fig. 1. Transition probability function of contexts *The hat* and *Not worn by a person* when $\alpha = 0$, the horizontal line at $\mu(\cdot, \cdot, \hat{p}) = 0.16$ shows the transition threshold used to identify atypical exemplars in table 1

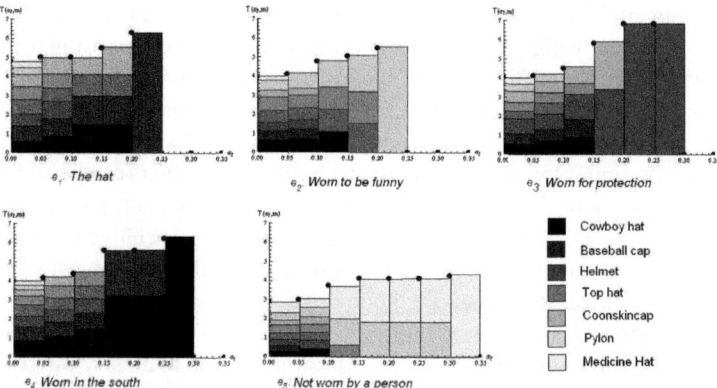

Fig. 2. Relevance of the contexts considered in the experiment, with respect to the threshold α

For example, consider the context e_5: *Not worn by a person* and the unit context e_1: *The hat*. We have $2.87 = T(e_5, 0) < T(e_1, 0) = 4.82$. But most of the contributions to $T(e_5, 0)$ come from the exemplars p_7: 'Toque' and p_8: 'Medicine Hat'. Indeed, $L(p_7, e_5)\mu(p_7, e_5, \hat{p}) + L(p_8, e_5)\mu(p_8, e_5, \hat{p}) = 2.46$. On the other hand, e_1 is the most typical context at zero threshold because many exemplars have a high Likert score. Thus, the values of its transitions probabilities $\mu(\cdot, e_1, \hat{p})$ are spread more homogeneously among the exemplars, leading to a flatter distribution with smaller probability values than the more typical exemplars of the e_5 distribution (see figure 1). If the threshold α is sufficiently high ($\alpha \geq 0.21$ in this case), $\mu_\alpha(\cdot, e_1, \hat{p})$ becomes the zero function because all the states in context e_1 are improbable for the threshold α, but context e_5 maintain their most probable states (p_7: 'Toque' and p_8: 'Medicine Hat'), because the transition probabilities of the states p_7 and p_8 are higher than α. Furthermore, the transition probabilities are amplified in the renormalized transition function $\mu_\alpha(\cdot, e_5, \hat{p})$ because $p_1, ..., p_6$ are improbable in context e_5 for the threshold $\alpha = 0.21$. This observation makes it possible to explain how we can use the transition threshold to gain a clearer picture of what is going on here.

These results reveal a dependency relationship between the threshold and the expected context typicality T. Figure 2 shows the function $T(e, \alpha)$ for different values of α for each context. What actually comes to mind depends both on alpha and on the context you are in, and Figure 2 expresses both of these. The top bar of each graph shows the relevance of the context for the corresponding value of alpha. The different coloured bars indicate which exemplars are available to transition to for the given value of the threshold α. We posit that the more different coloured bars there are, the greater the potential for entanglement of the different exemplars. The area of the filled box for a particular exemplar represents the transition probability with respect to the total size of the bar for the corresponding alpha. We considered the values of $\alpha \in \{0, 0.05, 0.1, 0.15, 0.2, 0.25, 0.3, 0.35\}$ to show how different exemplars remain able to be activated for different contexts, and how the probability distribution is affected by the renormalization. First, note that the expected typicality is an increasing function with respect to α until it reaches a maximal value that deactivates all exemplars. This is because the threshold is imposed to deactivate exemplars for which the transition probability is not sufficiently high, thus the remaining exemplars after imposing the threshold are those with higher transition probabilities. This implies that these remaining exemplars have comparatively higher typicality. Thus for the renormalized probability distribution, their expected typicality increases. Secondly, note that contexts $\{e_1, e_2, e_3, e_4\}$ are qualitatively similar for small values of α, i.e. all the exemplars can be activated with small probability values. However, the differences among the contexts are amplified as the threshold increases. This implies that in our model, an associative mode of thought permits activation of more exemplars at the cost of losing the meaningful specificity of the context. In contrast, in an analytic mode of thought, fewer exemplars are activated and they have higher transition probabilities due to the amplification of their probability values induced by the renormalization μ_α. Thus one is able to clearly differentiate the meaning of each context, at the cost of having less exemplars available for transition to.

Note that the threshold that makes no transition possible (all exemplars deactivated) varies with the context. The value required to deactivate all exemplars reflects the flatness of the probability distribution at $\alpha = 0$. The flatter the distribution, the smaller the value of α required to deactivate all exemplars. Indeed, in our model, context $e_1 = The$ hat requires the smallest threshold. This is because as e_1 gets flatter, the transition probabilities at $\alpha = 0$ have values close to the average probability $\mu^* = \frac{1}{8}$. For context e_5, the qualitative behavior with respect to α, i.e. the deactivation of certain exemplars as the threshold α increases, is the same as in the other contexts. However, context e_5 differs from other contexts in two important respects. First, e_5 is the only context that remains activated for exemplar p_8 :'Medicine Hat' for $\alpha > 0$, and is the only context that deactivates the exemplars p_1 :'Cowboy hat' and p_2 :'Baseball cap' for small values of α. Secondly, e_5 is the context that requires the largest threshold to deactivate all its exemplars. This is because e_5 has the most rugged distribution at $\alpha = 0$. Indeed, most of the transition probability at $\alpha = 0$ is concentrated on exemplars p_7 :'Pylon' and p_8 :'Medicine Hat'. These differences between e_5 and the rest of the contexts reflect the semantic opposition that context $e_5 = Not$ $worn$ by a $person$ has with the other contexts that state circumstances in which the concept HAT is elicited in a common-sense meaningful way.

Table 2. Types of contexts and the type of exemplars they have

$T(e)$	# typical exemplars	Context relevance at $\alpha = 0$	Type of exemplar
Large	Large	High	Very Representative
Medium	Large	Medium	Poorly representative
Medium	*Small*	*Low*	*Unexpected*
Small	Small	Low	Non-representative

6 Discussion and Future Directions

This paper builds on previous work that uses, SCOP, a quantum-inspired theory of concepts, and psychological data, to model conceptual structure, and specifically semantic relations between the different contexts that can influence a concept. Here we focus on how these contexts come into play in analytic versus associative thought. It is suggested that the notion of a transition threshold that shifts depending on the mode of thought, as well as newly defined notions of state and context expected typicality, are building blocks of a formal theory of creative thinking based on state transition probabilities in concepts. We posit that the more exemplars come to mind given a particular context and mode of thought, the greater the potential for entanglement of the different exemplars. The model is consistent with the occasional finding of unexpected meanings or interpretations of concepts. We propose that these new associations occur when a new context creates an unlikely set of new exemplars, which may potentially they exert quantum-like effects on one another. The paper also strengthens previous evidence that in order to account for the multiple meanings and flexible properties that concepts can assume, it is necessary to incorporate context into the concept definition.

The model developed here is small and idealized. In future research we plan to extend and generalize it. An interesting parameter that we have not yet explored is the sum of the expected typicality of a single exemplar with respect to the set of contexts. We believe that this can be interpreted as a measure of the *exemplar representativeness* given in Table 2. Much as the expected typicality of any given context is subject to change, unexpected exemplars could become more or less representative if the transition threshold changes. Further analysis could provide a richer description of this. Another interesting development is to study the structure of the transition probabilities when applying successive renormalizations induced by *sequences of thresholds* imposed to the concept structure. We could establish, straight from the data, a threshold-dependent hierarchy of pairs (p, e), that gives an account of the context-dependent *semantic distance* between exemplars. This could be used to model the characteristic, revealing, and sometimes surprising ways in which people make associations.

Acknowledgments. We are grateful for funding to Liane Gabora from the Social Sciences and Humanities Research Council of Canada and the Concerted Research Program of the Flemish Government of Belgium.

References

1. Piron, C.: Foundations of Quantum Physics. Reading, Benjamin (1976)
2. Aerts, D.: Being and change: foundations of a realistic operational formalism. In: Aerts, D., Czachor, M., Durt, T. (eds.) Probing the Structure of Quantum Mechanics: Nonlinearity, Nonlocality, Computation and Axiomatics. World Scientific, Singapore (2002)
3. Aerts, D., Gabora, L.: Contextualizing concepts using a mathematical generalization of the quantum formalism. J. Theor. Artif. Intell. 14, 327–358 (2002)
4. Aerts, D., Gabora, L.: A state-context-property model of concepts and their combinations i: The structure of the sets of contexts and properties. Kybernetes 34(1/2), 151–175 (2005)
5. Aerts, D., Gabora, L.: A state-context-property model of concepts and their combinations ii: A hilbert space representation. Kybernetes 34(1/2), 176–204 (2005)
6. Gabora, L., Aerts, D.: A model of the emergence and evolution of integrated worldviews. Journal of Mathematical Psychology 53, 434–451 (2009)
7. Ashby, F., Ell, S.: Stevens' handbook of experimental psychology: Methodology in experimental psychology, vol. 4. Wiley, New York (2002)
8. Evans, J., Frankish, K.: In two minds: Dual processes and beyond. Oxford University Press, New York (2009)
9. Hampton, J.: Inheritance of attributes in natural concept conjunctions. Memory & Cognition 15, 55–71 (1997)
10. Piaget, J.: The Language and Thought of the Child. Harcourt Brace, Kent UK (1926)
11. Gabora, L.: Cultural focus: A cognitive explanation for the cultural transition of the Middle/Upper Paleolithic. In: Proceedings of the 25th Annual Meeting of the Cognitive Science Society (2003)
12. Gabora, L.: Revenge of the 'neurds': Characterizing creative thought in terms of the structure and dynamics of human memory. Creativity Research Journal 22(1), 1–13 (2010)
13. Gabora, L., Rosch, E., Aerts, D.: Toward an ecological theory of concepts. Ecological Psychology 20(1), 84–116 (2008)
14. Guilford, P.: Creativity. American Psychologist 5, 444–454 (1950)
15. Finke, R., Ward, T., Smith, S.: Creative cognition: Theory, research and applications. MIT Press, Cambridge (1992)
16. Chaiken, S., Trope, Y.: Dual-process theories in social psychology. Guilford Press, New York (1999)
17. Freud, S.: An outline of psychoanalysis. Norton, New York (1949)
18. Russ, S.: Affect and creativity. Erlbaum, Hillsdale (1993)
19. Barsalou, L.W.: Context-independent and context-dependent information in concepts. Memory & Cognition 10, 82–93 (1982)
20. James, W.: The principles of psychology. Dover, New York (1890)

A Compositional Distributional Semantics, Two Concrete Constructions, and Some Experimental Evaluations

Mehrnoosh Sadrzadeh* and Edward Grefenstette

Department of Computer Science, University of Oxford, UK
{mehrnoosh.sadrzadeh,edward.grefenstette}@cs.ox.ac.uk

Abstract. We provide an overview of the hybrid compositional distributional model of meaning, developed in [6], which is based on the categorical methods also applied to the analysis of information flow in quantum protocols. The mathematical setting stipulates that the meaning of a sentence is a linear function of the tensor products of the meanings of its words. We provide concrete constructions for this definition and present techniques to build vector spaces for meaning vectors of words, as well as that of sentences. The applicability of these methods is demonstrated via a toy vector space as well as real data from the British National Corpus and two disambiguation experiments.

Keywords: Logic, Natural Language, Vector Spaces, Tensor Product, Composition, Distribution, Compact Categories, Pregroups.

1 Introduction

Words are the building blocks of sentences, yet the meaning of a sentence goes well beyond the meanings of its words. Indeed, while we do have dictionaries for words, we don't seem to need them to infer meanings of sentences. But where human beings seem comfortable doing this, machines fail to deliver. Automated search engines that perform well when queried by single words, fail to shine when it comes to search for meanings of phrases and sentences. Discovering the process of meaning assignment in natural language is among the most challenging as well as foundational questions of linguistics and computer science. The findings thereof will increase our understanding of cognition and intelligence and will also assist in applications to automating language-related tasks such as document search.

To date, the compositional type-logical [17,13] and the distributional vector space models [21,8] have provided two complementary partial solutions to the question. The logical approach is based on classic ideas from mathematical logic, mainly Frege's principle that meaning of a sentence can be derived from the relations of the words in it. The distributional model is more recent, it can be related to Wittgenstein's philosophy of 'meaning as use', whereby meanings of

---

* Support by EPSRC (grant EP/F042728/1) is gratefully acknowledged.

D. Song et al. (Eds.): QI 2011, LNCS 7052, pp. 35–47, 2011.

words can be determined from their context. The logical models have been the champions of the theory side, but in practice their distributional rivals have provided the best predictions.

In a cross-disciplinary approach, [6] used techniques from logic, category theory, and quantum information to develop a compositional distributional semantics that brought the above two models together. They developed a hybrid categorical model which paired contextual meaning with grammatical form and defined meaning of a string of words to be a function of the tensor product of the meanings of its words. As a result, meanings of sentences became vectors which lived in the same vector space and it became possible to measure their synonymity the same way lexical synonymity was measured in the distributional models. This sentence space was taken to be an abstract space and it was only shown how to instantiate it for the truth-functional meaning. Later [9] introduced a concrete construction using structured vector spaces and exemplified the application of logical methods, albeit only a toy vector space. In this paper we report on this and on a second construction which uses plain vector spaces. We also review results on implementing and evaluating the setting on real large scale data from the British National Corpus and two disambiguation experiments [10].

2   Sketching the Problem and a Hybrid Solution

To compute the meaning of a sentence consisting of n words, meanings of these words must *interact* with one another. In the logical models of meaning, this further interaction is represented in a function computed from the grammatical structure of the sentence, but meanings of words are empty entities. The grammatical structure is usually depicted as a parse-tree, for instance the parse-tree of the transitive sentence 'dogs chase cats' is as follows:

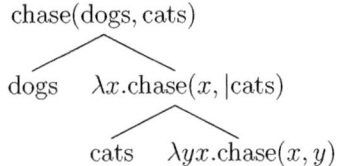

The function corresponding to this tree is based on a relational reading of the meaning of the verb 'chase', which makes the subject and the object interact with each other via the relation of *chasing*. This methodology is used to translate sentences of natural language into logical formulae, then use computer-aided automation tools to reason about them [2]. The major drawback is that the result can only deal with truth or falsity as the meaning of a sentence and does poorly on lexical semantics, hence do not perform well on language tasks such as search.

The vector space model, on the other hand, dismisses the further interaction and is solely based on lexical semantics. These are obtained in an operational way, best described by a frequently cited quotation due to Firth [8] that "You shall know a word by the company it keeps.". For instance, beer and sherry are both drinks, alcoholic, and often make you drunk. These facts are reflected in the text: words 'beer' and 'sherry' occur close to 'drink', 'alcoholic' and 'drunk'. Hence meanings of words can be encoded as vectors in a highly dimensional space of context words. The raw weight in each base is related to the number of times the word has appeared close (in an n-word window) to that base. This setting offers geometric means to reason about meaning similarity, e.g. via the cosine of the angle between the vectors. Computational models along these lines have been built using large vector spaces (tens of thousands of basis vectors) and large bodies of text (up to a billion words) [7]. These models have responded well to language processing tasks such as word sense discrimination, thesaurus construction, and document retrieval [11,21]. Their major drawback is their non-compositional nature: they ignore the grammatical structure and logical words, hence cannot compute (in the same efficient way that they do for words) meanings of phrases and sentences.

The key idea behind the approach of [6] is to import the compositional element of the logical approaches into the vector space models by making the grammar of the sentence *act* on, hence relate, its word vectors. The trouble is that it does not make so much sense to 'make a parse tree act on vectors'. Some higher order mathematics, in this case category theory, is needed to encode the grammar of a sentence into a morphism compatible with vector spaces[1]. These morphisms turn out to be the grammatical reductions of a type-logic called a Lambek pregroup [13]. Pregroups and vector spaces both have a *compact* categorical structural. The grammatical morphism of a pregroup can be transformed into a linear map that acts on vectors. Meanings of sentences become vectors whose angles reflect similarity. Hence, at least theoretically, one should be able to build sentence vectors and compare their synonymity, in exactly the same way as measuring synonymity for words.

The pragmatic interpretation of this abstract idea is as follows. In the vector space models, one has a meaning vector for each word, $\overrightarrow{\text{dogs}}, \overrightarrow{\text{chase}}, \overrightarrow{\text{cats}}$. The logical recipe tells us to *apply* the meaning of verb to the meanings of subject and object. But how can a vector *apply* to other vectors? If we strip the vectors off the extra information provided in their basis and look at them as mere sets of weights, then we can apply them to each other by taking their point-wise sum or product. But these operations are commutative, whereas meaning is not. Hence this will equalize meaning of any combination of words, even with the non-grammatical combinations such as 'dogs cats chase'. The proposed solution above implies that one needs to have different levels of meaning for words with different functionalities. This is similar to the logical models whereby verbs are relations and nouns are atomic sets. So verb vectors should be built differently from noun vectors, for instance as matrices that relate and act on the atomic noun vectors. The general information, as to which words should be matrices and which atomic vectors, is in fact encoded in the type-logical representation of

---

[1] A similar passage had to be made in other type-logics to turn the parse-trees into lambda terms, compatible with sets and relations.

the grammar. That is why the grammatical structure of the sentence is a good candidate for the process that relates its word vectors.

In a nutshell, pregroup types are either atomic or compound. Atomic types can be simple (*e.g.* n for noun phrases, s for statements) or left/right superscripted— referred to as adjoint types (*e.g.* n^r and n^l). An example of a compound type is that of a verb $n^r s n^l$. The superscripted types express that the verb is a relation with two arguments of type n, which have to occur to the <u>r</u>ight and to the <u>l</u>eft of it, and that it outputs an argument of the type s. A transitive sentence is typed as shown below.

$$\begin{array}{ccc} \text{dogs} & \text{chase} & \text{cats.} \\ n & n^r\, s\, n^l & n \end{array}$$

Here, the verb interacts with the subject and object via the underlying wire cups, then produces a sentence via the outgoing line. These interactions happen in real time. The type-logical analysis assigns type n to 'dogs' and 'cats', for a noun phrase, and the type $n^r s n^l$ to 'chase' for a verb, the superscripted types n^r and n^l express the fact that the verb is a function with two arguments of type n, which have to occur to the *right* and *left* of it. The reduction computation is $nn^r s n^l \leq 1s1 = s$, each type n cancels out with its right *adjoint* n^r from the right, i.e. $nn^r \leq 1$ and its left adjoint n^l from the left, i.e. $n^l n \leq 1$, and 1 is the unit of concatenation $1n = n1 = n$. The algebra advocates a linear method of parsing: a sentence is analyzed as it is heard, i.e. word by word, rather than by first buffering the entire string then re-adjusting it as necessary on a tree. It's been argued that the brain works in this one-dimensional linear (rather than two-dimensional tree) manner [13].

According to [6] and based on a general completeness theorem between compact categories, wire diagrams, and vector spaces, meaning of sentences can be canonically reduced to linear algebraic formulae, for example the following is the meaning vector of our transitive sentence:

$$\overrightarrow{\text{dogs chase cats}} = (f)\left(\overrightarrow{\text{dogs}} \otimes \overrightarrow{\text{chase}} \otimes \overrightarrow{\text{cats}}\right)$$

Here f is the linear map that encodes the grammatical structure. The categorical morphism corresponding to it is denoted by the tensor product of 3 components: $\epsilon_V \otimes 1_S \otimes \epsilon_W$, where V and W are subject and object spaces, S is the sentence space, the ϵ's are the cups, and 1_S is the straight line in the diagram. The cups stand for taking inner products, which when done with the basis vectors imitate substitution. The straight line stands for the identity map that does nothing. By the rules of the category, the above equation reduces to the following linear algebraic formula with lower dimensions, hence the dimensional explosion problem for tensor products is avoided:

$$\sum_{itj} C^{\text{chase}}_{itj} \langle \overrightarrow{\text{dogs}} \mid \overrightarrow{v_i} \rangle \overrightarrow{s_t} \langle \overrightarrow{w_j} \mid \overrightarrow{\text{cats}} \rangle \in S$$

In the above equation, $\overrightarrow{v_i}, \overrightarrow{w_j}$ are basis vectors of V and W. The meaning of the verb becomes a superposition, represented as a linear map. The inner product $\langle \overrightarrow{\text{dogs}} | \overrightarrow{v_i} \rangle$ substitutes the weights of $\overrightarrow{\text{dogs}}$ into the first argument place of the verb (similarly for object and second argument place) and results in producing a vector for the meaning of the sentence. These vectors live in sentence spaces S, for which $\overrightarrow{s_t}$ is a base vector. The degree of synonymity of sentences is obtained by taking the cosine measure of their vectors. S is an abstract space, it needs to be instantiated to provide concrete meanings and synonymity measures. For instance, a truth-theoretic model is obtained by taking the sentence space S to be the 2-dimensional space with basis vector true $|1\rangle$ and false $|0\rangle$. This is done by using the weighting factor C_{itj}^{chase} to define a model-theoretic meaning for the verb as follows:

$$C_{itj}^{\text{chase}} \overrightarrow{s_t} = \begin{cases} |1\rangle & chase(v_i, w_j) = \text{true}, \\ |0\rangle & o.w. \end{cases}$$

The definition of our meaning map ensures that this value propagates to the meaning of the whole sentence. So $chase(\text{dogs}, \text{cats})$ becomes true whenever 'dogs chase cats' is true and false otherwise.

3 Two Concrete Constructions for Sentence Spaces

The above construction is based on the assumptions that $\overrightarrow{\text{dogs}}$ is a base of V and that $\overrightarrow{\text{cats}}$ is a base of W. In other words, we assume that V is the vector space spanned by the set of all men and W is the vector space spanned by the set of all women. This is not the usual construction in the distributional models. In what follows we present two concrete constructions for these, which will then yield a construction for the sentence space. In both of these approaches V and W will be the same vector space, which we will denote by N.

3.1 Structured Vector Spaces and a Toy Corpus

We take N to be a *structured vector space*, as in [11]. The bases of N are annotated by 'properties' obtained by combining dependency relations with nouns, verbs and adjectives. For example, basis vectors might be associated with properties such as "arg-fluffy", denoting the argument of the adjective fluffy, "subj-chase" denoting the subject of the verb chase, "obj-buy" denoting the object of the verb buy, and so on. We construct the vector for a noun by counting how many times in the corpus a word has been the argument of 'fluffy', the subject of 'chase', the object of 'buy', and so on.

For transitive sentences, we take the sentence space S to be $N \otimes N$, so its bases are of the form $\overrightarrow{s_t} = (\overrightarrow{n_i}, \overrightarrow{n_j})$. The intuition is that, for a transitive verb, the meaning of a sentence is determined by the meaning of the verb together with its subject and object. The verb vectors $C_{itj}^{\text{verb}}(\overrightarrow{n_i}, \overrightarrow{n_j})$ are built by counting how many times a word that is n_i (e.g. has the property of being fluffy) has been subject of the verb and a word that is n_j (e.g. has the property that it's bought) has been its object, where the counts are moderated by the extent to which the subject and object exemplify each property (e.g. *how fluffy* the subject is). To give a rough paraphrase of the intuition behind this approach, the meaning of "dog chases cat" is given by: the extent to which a dog is fluffy and a cat is something that is bought (for the $N \otimes N$ property pair "arg-fluffy" and "obj-buy"), and the extent to which fluffy things *chase* things that are bought (accounting for the meaning of the verb for this particular property pair); plus the extent to which a dog is something that runs and a cat is something that is cute (for the $N \otimes N$ pair "subj-run" and "arg-cute"), and the extent to which things that run *chase* things that are cute (accounting for the meaning of the verb for this particular property pair); and so on for all noun property pairs.

For sentences with intransitive verbs, the sentence space suffices to be just N. To compare the meaning of a transitive sentence with an intransitive one, we embed the meaning of the latter from N into the former $N \otimes N$, by taking $\overrightarrow{\varepsilon_n}$ (the 'object' of an intransitive verb) to be $\sum_i \overrightarrow{n_i}$, i.e. the superposition of all basis vectors of N. A similar method is used while dealing with sentences with ditransitive verbs, where the sentence space will be $N \otimes N \otimes N$, since these verbs have three arguments. Transitive and intransitive sentences are then embedded in this bigger space, using the same embedding described above.

Adjectives are dealt with in a similar way. We give them the syntactic type nn^l and build their vectors in $N \otimes N$. The syntactic reduction $nn^l n \to n$ associated with applying an adjective to a noun gives us the map $1_N \otimes \epsilon_N$ by which we semantically compose an adjective with a noun, as follows:

$$\overrightarrow{\text{adjective noun}} = (1_N \otimes \epsilon_N)(\overrightarrow{\text{adj}} \otimes \overrightarrow{\text{noun}}) = \sum_{ij} C_{ij}^{\text{adj}} \overrightarrow{n_i} \langle \overrightarrow{n_j} \mid \overrightarrow{\text{noun}} \rangle$$

We can view the C_{ij}^{adj} counts as determining what sorts of properties the arguments of a particular adjective typically have (e.g. arg-red, arg-colourful for the adjective "red").

As an example, consider a hypothetical vector space with bases 'arg-fluffy', 'arg-ferocious', 'obj-buys', 'arg-shrewd', 'arg-valuable', with vectors for 'bankers', 'cats', 'dogs', 'stock', and 'kittens'.

		bankers	cats	dogs	stock	kittens
1	arg-fluffy	0	7	3	0	2
2	arg-ferocious	4	1	6	0	0
3	obj-buys	0	4	2	7	0
4	arg-shrewd	6	3	1	0	1
5	arg-valuable	0	1	2	8	0

Since in the method proposed above, $C_{itj}^{\text{verb}} = 0$ if $\overrightarrow{s_t} \neq (\overrightarrow{n_i}, \overrightarrow{n_j})$, we can simplify the weight matrices for transitive verbs to two dimensional C_{ij}^{verb} matrices as shown below, where C_{ij}^{verb} corresponds to the number of times the verb has a subject with attribute n_i and an object with attribute n_j. For example, the matrix below encodes the fact that something ferocious ($i = 2$) chases something fluffy ($j = 1$) seven times in the hypothetical corpus from which we might have obtained these distributions.

$$C^{\text{chase}} = \begin{bmatrix} 1 & 0 & 0 & 0 & 0 \\ 7 & 1 & 2 & 3 & 1 \\ 0 & 0 & 0 & 0 & 0 \\ 2 & 0 & 1 & 0 & 1 \\ 1 & 0 & 0 & 0 & 0 \end{bmatrix}$$

Once we have built matrices for verbs, we are able to follow the categorical procedure and automatically build vectors for sentences, then perform sentence comparisons. The comparison is done in the same way as for lexical semantics, i.e. by taking the inner product of the vectors of two sentences and normalizing it by the product of their lengths. For example the following shows a high similarity

$$cos(\overrightarrow{\text{dogs chase cats}}, \overrightarrow{\text{dogs pursue kittens}}) = \frac{\langle \overrightarrow{\text{dogs chase cats}} \mid \overrightarrow{\text{dogs pursue kittens}} \rangle}{\mid \overrightarrow{\text{dogs chase cats}} \mid \times \mid \overrightarrow{\text{dogs pursue kittens}} \mid} =$$

$$\frac{\left\langle \left(\sum_{itj} C_{itj}^{\text{chase}} \langle \overrightarrow{\text{dogs}} \mid \overrightarrow{n_i} \rangle \overrightarrow{s_t} \langle \overrightarrow{n_j} \mid \overrightarrow{\text{cats}} \rangle \right) \mid \left(\sum_{itj} C_{itj}^{\text{pursue}} \langle \overrightarrow{\text{dogs}} \mid \overrightarrow{n_i} \rangle \overrightarrow{s_t} \langle \overrightarrow{n_j} \mid \overrightarrow{\text{kittens}} \rangle \right) \right\rangle}{\mid \overrightarrow{\text{dogs chase cats}} \mid \times \mid \overrightarrow{\text{dogs pursue kittens}} \mid}$$

$$= \frac{\sum_{itj} C_{itj}^{\text{chase}} C_{itj}^{\text{pursue}} \langle \overrightarrow{\text{dogs}} \mid \overrightarrow{n_i} \rangle \langle \overrightarrow{\text{dogs}} \mid \overrightarrow{n_i} \rangle \langle \overrightarrow{n_j} \mid \overrightarrow{\text{cats}} \rangle \langle \overrightarrow{n_j} \mid \overrightarrow{\text{kittens}} \rangle}{\mid \overrightarrow{\text{dogs chase cats}} \mid \times \mid \overrightarrow{\text{dogs pursue kittens}} \mid} = 0.979$$

A similar computation will provide us with the following, demonstrating a low similarity

$$cos(\langle \overrightarrow{\text{dogs chase cats}} \mid \overrightarrow{\text{bankers sell stock}} \rangle) = 0.042$$

The construction for adjective matrices are similar: we stipulate the C_{ij}^{adj} matrices by hand and eliminate all cases where $i \neq j$ since $C_{ij} = 0$, hence these become one dimensional matrices. Here is an example

$$C^{\text{fluffy}} = [9\ 3\ 4\ 2\ 2]$$

Vectors for 'adjective noun' clauses are computed similarly and are used to compute the following similarity measures:

$$cosine(\overrightarrow{\text{fluffy\ \ dog}}, \overrightarrow{\text{shrewd\ \ banker}}) = 0.389$$
$$cosine(\overrightarrow{\text{fluffy\ \ cat}}, \overrightarrow{\text{valuable\ \ stock}}) = 0.184$$

These calculations carry over to sentences which contain the 'adjective noun' clauses. For instance, we obtain an even lower similarity measure between the following sentences:

$$cosine(\overrightarrow{\text{fluffy dogs chase fluffy cats}}, \overrightarrow{\text{shrewd bankers sell valuable stock}}) = 0.016$$

Other constructs such as prepositional phrases and adverbs are treated similarly, see [9].

3.2 Plain Vector Spaces and the BNC

The above concrete example is fine grained, but involves complex constructions which are time and space costly when implemented. To be able to evaluate the setting against real large scale data, we simplified it by taking N to be a plain vector spaces whose bases are words, without annotations. The weighting factor C_{ij}^{verb} is determined in the same as above, but this time by just counting co-occurence rather than being arguments of syntactic roles. More precisely, this weight is determined by the number of times the subjects of the verb have co-occured with the base $\vec{n}_i$. In the previous construction we went beyond co-occurence and required that the subject (similarly for the object) should be in a certain relation with the verb, for instance if $\vec{n}_i$ was 'arg-fluffly', the subject had to be an argument of fluffy, where as here we instead have $\vec{n}_i = $ 'fluffy', and the subject has to co-occure with 'fluffy' rather than being directly modified by it.

The procedure for computing these weights for the case of transitive sentences is as follows: first browse the corpus to find all occurrences of the verb in question, suppose it has occurred as a transitive verb in k sentences. For each sentence determine the subject and the object of the verb. Build vectors for each of these using the usual distributional method. Multiply their weights on all permutations of their coordinates and then take the sum of each such multiplication across each of the k sentences. Linear algebraically, this is just the sum of the Kronecker products of the vectors of subjects and objects:

$$\overrightarrow{verb} \;=\; \sum_k \left(\overrightarrow{sub} \otimes \overrightarrow{obj} \right)_k$$

Recall that given a vector space A with basis $\{\vec{n}_i\}_i$, the Kronecker product of two vectors $\vec{v} = \sum_i c_i^a \vec{n}_i$ and $\vec{w} = \sum_i c_i^b \vec{n}_i$ is defined as follows:

$$\vec{v} \otimes \vec{w} = \sum_{ij} c_i^a c_j^b \left(\vec{n}_i \otimes \vec{n}_j \right)$$

As an example, we worked with the British National Corpus (BNC) which has about 6 million sentences. We built noun vectors and computed matrices for intransitive verbs, transitive verbs, and adjectives. For instance, consider N to be the space with four basis vectors 'far', 'room', 'scientific', and 'elect'; the (TF/IDF) values for vectors of the four nouns 'table', 'map', 'result', and 'location' are shown below.

A section of the matrix of the transitive verb 'show' is represented below.

As a sample computation, suppose the verb 'show' only appears in two sentences in the corpuse: 'the map showed the location' and 'the table showed the result'. The weight c_{12} for the base i.e. $(\overrightarrow{far}, \overrightarrow{far})$ is computed by multiplying weights of 'table' and 'result' on $\overrightarrow{far}$, i.e. 6.6×7, multiplying weights of 'map' and 'location' on $\overrightarrow{far}$, i.e. 5.6×5.9 then adding these $46.2 + 33.04$ and obtaining the total weight 79.24.

The computations for building vectors for sentences and other phrases are the same as in the case for structured vector spaces. The matrix of a transitive verb has

Table 1. Sample noun vectors from the BNC

i	$\overrightarrow{n_i}$	table	map	result	location
1	far	6.6	5.6	7	5.9
2	room	27	7.4	0.99	7.3
3	scientific	0	5.4	13	6.1
4	elect	0	0	4.2	0

Table 2. Sample verb matix from the BNC

	far	room	scientific	elect
far	79.24	47.41	119.96	27.72
room	232.66	80.75	396.14	113.2
scientific	32.94	31.86	32.94	0
elect	0	0	0	0

2 dimensions since it takes as input two arguments. The same method is applied to build matrices for ditransitive verbs, which will have 3 dimensions, and intransitive verbs, as well as adjectives and adverbs, which will be of 1 dimension each.

4 Evaluation and Experiments

We evaluated our second concrete method on a disambiguation task and performed two experiments [10]. The general idea behind this disambiguation task is that some verbs have different meanings and the context in which they appear is used to disambiguate them. For instance the verb 'show' can mean 'express' in the context 'the table showed the result' or it can mean 'picture', in the context 'the map showed the location'. Hence if we build meaning vectors for these sentences compositionally, the degrees of synonymity of the sentences can be used to disambiguate the meaning of the verb in that sentence. Suppose a verb has two meanings and it has occurred in two sentences. Then if in both of these sentences it has its meaning number 1, the two sentences will have a high degree of synonymity, whereas if in one sentence the verb has its meaning number 1 and in the other its meaning number 2, the sentences will have a lower degree of synonymity. For instance, 'the table showed the result' and 'the table expressed the result', have a hight degree of synonymity and similarly for 'the map showed the location' and 'the map pictured the location'. This degree decreases for the two sentences 'the table showed the result' and 'the table pictured the result'. We used our second concrete construction to implement this task.

The data set for our first experiment was developed by [16] and had 120 sentence pairs. These were all intransitive sentences. We compared the results of our method with composition operations implemented by [16], these included addition, multiplication, and a combination of two using weights. The best results were obtained by the multiplication operator. Our method provided slightly better results. However, the context provided by intransitive sentences is just one word, hence the results do not showcase the compositional abilities of our method. In particular, in such a small context, our method and the multiplication method became very similar, hence the similarity of results did not surprise us. There is nevertheless two major differences: our method respects the grammatical structure of the sentences (whereas the multiplication operation does not) and in our method the vector of the verb is computed differently from the vectors of the nouns: as a relation and via a second order construction.

For the second experiment, we developed a data set of transitive sentences. We first picked 10 transitive verbs from the most occurring verbs of the BNC, each verb has at least two different non-overlapping meanings. These were retrieved using the JCN (Jiang Conrath) information content synonymity measure of WordNet. The above example for 'show' and its two meanings 'express' and 'picture' is one such example. For each such verb, e.g. 'show', we retrieved 10 sentences which contained them (as verbs) from the BNC. An example of such a sentence is 'the table showed the result'. We then substituted in each sentence each of the two meanings of the verb, for instance 'the table expressed the result' and 'the table pictured the result'. This provided us with 200 pairs of sentences and we used the plain method described above to build vectors for each sentence and compute the cosine of each pair. A sample of these pairs is provided below.

In order to judge the performance of our method, we followed guidelines from [16]. We distributed our data set among 25 volunteers who were asked to rank each pair based on how similar they thought they were. The ranking was between 1 and 7, where 1 was almost dissimilar and 7 almost identical. Each pair was also given a HIGH or LOW classification by us. The correlation of the model's similarity judgements with the human judgements was calculated using Spearman's ρ, a metric which is deemed to be more scrupulous and ultimately that by which models should be ranked. It is assumed that inter-annotator agreement provides the theoretical maximum ρ for any model for this experiment, and that taking the cosine measure of the verb vectors while ignoring the noun was taken as the baseline.

The results for the models evaluated against the both datasets are presented below. The additive and multiplicative operations are applications of vector addition and multiplication; Kintsch is a combination of the two, obtained by multiplying the word vectors by certain weighting constants and then adding them, for details please see [16]. The *Baseline* is from a non-compositional approach, obtained by only comparing vectors of verbs of the sentences and ignoring their

Table 3. Sample sentence pairs from the second experiment dataset

	Sentence 1	Sentence 2
1	table show result	table express result
2	table show result	table picture result
3	map show location	map picture location
4	map show location	map express location
5	child show interest	child picture interest
6	child show interest	child express interest

Table 4. Results of the 1st and 2nd compositional disambiguation experiments

Model	High	Low	ρ
Baseline	0.27	0.26	0.08
Add	0.59	0.59	0.04
Kintsch	0.47	0.45	0.09
Multiply	0.42	0.28	0.17
Categorical	**0.84**	**0.79**	**0.17**
UpperBound	4.94	3.25	0.40

Model	High	Low	ρ
Baseline	0.47	0.44	0.16
Add	0.90	0.90	0.05
Multiply	0.67	0.59	0.17
Categorical	**0.73**	**0.72**	**0.21**
UpperBound	4.80	2.49	0.62

subjects and objects. The *UpperBound* is the summary of the human ratings, also known as inter-annotator agreement.

According to the literature (e.g. see [16]), the main measure of success is demonstrated by the ρ column. By this measure in the second experiment our method outperforms the other two with a much better margin than that in the first experiment. The High (similarly Low) columns are the average score that High (Low) similarity sentences (as decided by us) get by the program. These are not very indicative, as the difference between high mean and the low mean of the categorical model is much smaller than that of the both the baseline model and multiplicative model, despite better alignment with annotator judgements.

The data set of the first experiment has a very simple syntactic structure where the context around the verb is just its subject. As a result, in practice the categorical method becomes very similar to the multiplicative one and the similar outcomes should not surprise us. The second experiment, on the other hand, has more syntactic structure, thereby our categorical shows an increase in alignment with human judgements. Finally, the increase of ρ from the first experiment to the second reflects the compositionality of our model: its performance increases with the increase in syntactic complexity. Based on this, we would like to believe that more complex datasets and experiments which for example include adjectives and adverbs shall lead to even better results.

5 Conclusion and Future Work

We have provided a brief overview of the categorical compositional distributional model of meaning as developed in [6]. This combines the logical and vector space models using the setting of compact closed categories and their diagrammatic toolkit and based on ideas presented in [5] on the use of tensor product as a meaning composition operator. We go over two concrete constructions of the setting, show examples of one construction on a toy vector space and implement the other construction on the real data from the BNC. The latter is evaluated on a disambiguation task on two experiments: for intransitive verbs from [16] and for transitive verbs developed by us. The categorical model slightly improves the results of the first experiment and betters them in the second one.

To draw a closer connection with the subject area of the workshop, we would like to recall that sentences of natural language are compound systems, whose meanings exceed the meanings of their parts. Compound systems are a phenomena studied by many sciences, findings thereof should as well provide valuable insights for natural language processing. In fact, some of the above observations and previous results were led by the use of compact categories in compound quantum systems [1]. The caps that connect subject and verb from afar are used to model nonlocal correlations in entangled Bell states; meanings of verbs are represented as superposed states that let the information flow between their subjects and objects and further act on it. Even on the level of single quantum systems, there are similarities to the distributional meanings of words: both are modeled using vector spaces. Motivated by this [19,22] have used the methods of quantum logic to provide logical and geometric structures for information retrieval and have also obtained better results in practice. We hope and aim to study the modular extension of the quantum logic methods to tensor spaces of our approach. There are other approaches to natural language processing that use compound quantum systems but which do not focus on distributional models, for example see [4].

Other areas of future work include creating and running more complex experiments that involve adjectives and adverbs, working with larger corpora such as the WaCKy, and interpreting stop words such as relative pronouns *who, which,* conjunctives *and, or,* and quantifiers *every, some.*

References

1. Abramsky, S., Coecke, B.: A categorical semantics of quantum protocols. In: Proceedings of the 19th Annual IEEE Symposium on Logic in Computer Science (2004)
2. Alshawi, H. (ed.): The Core Language Engine. MIT Press, Cambridge (1992)
3. Baroni, M., Zamparelli, R.: Nouns are vectors, adjectives are matrices. In: Proceedings of Conference on Empirical Methods in Natural Language Processing, EMNLP (2010)
4. Bruza, P., Kitto, K., Nelson, D.L., McEvoy, C.L.: Entangling words and meaning. In: Proceedings of AAAI Spring Symposium on Quantum Interaction. Oxford University, College Publications (2008)
5. Clark, S., Pulman, S.: Combining Symbolic and Distributional Models of Meaning. In: Proceedings of AAAI Spring Symposium on Quantum Interaction. Standord University, AAAI Press (2007)
6. Coecke, B., Sadrzadeh, M., Clark, S.: Mathematical Foundations for Distributed Compositional Model of Meaning. In: van Benthem, J., Moortgat, M., Buszkowski, W. (eds.) Lambek Festschrift. Linguistic Analysis, vol. 36, pp. 345–384 (2010); arXiv:1003.4394v1 [cs.CL]
7. Curran, J.: From Distributional to Semantic Similarity. PhD Thesis, University of Edinburgh (2004)
8. Firth, J.R.: A synopsis of linguistic theory 1930-1955. Studies in Linguistic Analysis (1957)

 9. Grefenstette, E., Sadrzadeh, M., Clark, S., Coecke, B., Pulman, S.: Concrete Compositional Sentence Spaces for a Compositional Distributional Model of Meaning. In: International Conference on Computational Semantics (IWCS 2011), Oxford (2011); arXiv:1101.0309v1 [cs.CL]
10. Grefenstette, E., Sadrzadeh, M.: Experimental Support for a Categorical Compositional Distributional Model of Meaning. In: Empirical Methods in Natural Language Processing (EMNLP 2011), Edinburgh (2011)
11. Grefenstette, G.: Explorations in Automatic Thesaurus Discovery. Kluwer, Dordrecht (1994)
12. Guevara, E.: A Regression Model of Adjective-Noun Compositionality in Distributional Semantics. In: Proceedings of the ACL GEMS Workshop (2010)
13. Lambek, J.: From Word to Sentence. Polimetrica, Milan (2008)
14. Landauer, T., Dumais, S.: A solution to Platos problem: The latent semantic analysis theory of acquisition, induction, and representation of knowledge. Psychological Review (2008)
15. Manning, C.D., Raghavan, P., Schütze, H.: Introduction to information retrieval. Cambridge University Press, Cambridge (2008)
16. Mitchell, J., Lapata, M.: Vector-based models of semantic composition. In: Proceedings of the 46th Annual Meeting of the Association for Computational Linguistics, pp. 236–244 (2008)
17. Montague, R.: English as a formal language. Formal Philosophy, 189–223 (1974)
18. Nivre, J.: An efficient algorithm for projective dependency parsing. In: Proceedings of the 8th International Workshop on Parsing Technologies, IWPT (2003)
19. van Rijsbergen, K.: The Geometry of Information Retrieval. Cambridge University Press, Cambridge (2004)
20. Saffron, J., Newport, E., Asling, R.: Word Segmentation: The role of distributional cues. Journal of Memory and Language 35, 606–621 (1999)
21. Schuetze, H.: Automatic Word Sense Discrimination. Computational Linguistics 24, 97–123 (1998)
22. Widdows, D.: Geometry and Meaning. University of Chicago Press, Chicago (2005)

Finding Schizophrenia's Prozac
Emergent Relational Similarity in Predication Space

Trevor Cohen[1], Dominic Widdows[2], Roger Schvaneveldt[3],
and Thomas C. Rindflesch[4]

[1] University of Texas Health Science Center at Houston
[2] Google, inc.
[3] Arizona State University
[4] National Library of Medicine

Abstract. In this paper, we investigate the ability of the Predication-based Semantic Indexing (PSI) approach, which incorporates both symbolic and distributional information, to support inference on the basis of structural similarity. For example, given a pair of related concepts prozac:depression, we attempt to identify concepts that relate to a third concept, such as schizophrenia in the same way. A novel PSI implementation based on Kanerva's Binary Spatter Code is developed, and evaluated on over 100,000 searches across 180,285 unique concepts and multiple typed relations. PSI is shown to retrieve with accuracy concepts on the basis of shared single and paired relations, given either a single strong example pair, or the superposition of a set of weaker examples. Search space size is identical for single and double relations, providing an efficient means to direct search across predicate paths for the purpose of literature-based discovery.

Keywords: Distributional Semantics, Vector Symbolic Architectures, Literature-based Discovery, Abductive Reasoning.

1 Introduction

This paper presents new results that demonstrate ways in which high-dimensional vector representations can be used to model proportional analogies such as "prozac is to depression as what is to schizophrenia?" Our approach is based on our earlier "Logical Leaps" work [1], and Kanerva's work on hyperdimensional computing and analogical mapping [2] (both presented at Quantum Informatics, 2010). This approach depends upon being able to represent concepts as high-dimensional vectors, and relationships between concepts as mathematical operations on these vectors. Such operations include composition of vectors using product and superposition operations, and the selection of nearby pure concepts from a superposed or product state. The work is part of the family of generalized quantum methods currently being explored: basic concepts are analogous to pure states; superposition and product operations give rise to compound concepts analogous to mixed and entangled states; and the selection of a nearby known concept from a product state is analogous to quantization or quantum collapse. A notable departure from traditional quantum mechanics is our use of real and binary vectors, instead of complex vectors. This departure is not novel and is an oft-understated discrepancy of approaches: for many years the information retrieval and machine learning communities have used real-valued vectors; Kanerva's work uses binary-valued vectors

D. Song et al. (Eds.): QI 2011, LNCS 7052, pp. 48–59, 2011.

as examples [2]; and traditional quantum mechanics almost exclusively used complex Hilbert spaces, as have emerging approaches to information retrieval [3] and distributional semantics [4]. We mention this at the outset as perhaps one of the key senses in which "generalized quantum" models should be thought of as generalizations, not applications, of quantum physics.

2   Background

The "Logical Leaps" approach is an extension of our previous work in the domain of literature-based discovery [5], in which we evaluated the ability of various scalable models of distributional semantics to generate *indirect inferences* [6], meaningful connections between terms that do not co-occur in any document in a given corpus. Connections of this sort are fundamental to Swanson's model of literature-based discovery [7], which emerged from the serendipitous discovery of a therapeutically useful [8] connection between Raynaud's Syndrome (reduced blood flow in the extremities) and fish oils. This connection was based on the bridging concept "blood viscosity": fish oil can decrease blood viscosity thus increasing blood flow. Swanson's method can be seen as an example of abductive reasoning, hypothesis generation as proposed by Peirce (see [9]), and provides the basis for several computer models that aim to facilitate discovery [10], [11]. As an alternative to stepwise exploration of the vast search space of possible bridging concepts and discoveries, distributional approaches such as Latent Semantic Analysis [6], Random Indexing (RI) [12] and others have been applied to infer meaningful indirect connections between terms without identifying a bridging concept [13], [14], [5]. In contrast to these approaches, which are based on general association strength, "Logical Leaps" are derived from a vector space in which both the target and the type of a relation to a concept are encoded into its vector representation. This has been achieved using Predication-based Semantic Indexing (PSI) [15], a variant of RI that uses permutation of sparse random vectors to encode relationships (such as TREATS) between concepts into a high-dimensional vector space. In this paper, we attempt to direct searches in PSI space by specifying predicate paths using a pair of example concepts. We achieve this end with an alternative implementation of PSI based on Kanerva's Binary Spatter Code which we introduce in the following section.

3   Mathematical Structure and Methods

The methods in this paper all use high-dimensional vectors to represent concepts. There are many ways of generating such representations. Ours is based upon the RI paradigm using terminology as described in [5], in which *semantic vectors* are built as superpositions of randomly generated *elemental vectors*, derived by training over a corpus of documents. Throughout this paper we will write $E(X)$ and $S(X)$ for the elemental and semantic vectors associated with the concept X. In addition to concept vectors, we introduce vectors for relations. For example, $E(R)$ would denote the elemental vector for the relation R. Many relationships are directional, and we will use R_{inv} to denote the inverse of R, so that A R B and B R_{inv} A carry the same external meaning (though they may in some cases be represented by different vectors).

Kanerva's Binary Spatter Code [16] provides the means to encode typed relations into a high-dimensional binary vector space. The Spatter Code is one of a group of representational approaches collectively known as Vector Symbolic Architectures [17] (VSAs), which originated from Smolensky's tensor product based approach [18], and include Holographic Reduced Representations (HRRs) [19] amongst others. VSAs differ from earlier connectionist representations as they allow for the encoding of typed relations and nested compositional structure. Most of the definitions given below work for VSAs in general. However, we make particular use of VSAs with binary-valued vectors and component-wise exclusive or (XOR) as the binding operation: this has the special property of being its own inverse, which the reader should not assume for other implementations.

The primary operations facilitated by VSAs are *binding* and *bundling*. Binding is a multiplication-like operator through which two vectors are combined to form a third vector C that is dissimilar from either of its component vectors A and B. We will use the symbol "$\otimes$" for binding, and the symbol "$\oslash$" for the inverse of binding throughout this paper. Be aware that binding may have different implementations in different models, and is not meant to be identified with the tensor product. It is important that this operator be invertible: if $C = A \otimes B$, then $A \oslash C = A \oslash (A \otimes B) = B$. In some models, this recovery may be approximate, but the robust nature of the representation guarantees that $A \oslash C$ is similar enough to B that B can easily be recognized as the best candidate for $A \oslash C$ in the original set of concepts. Thus the invertible nature of the bind operator facilitates the retrieval of information encoded during the binding process. While this operator varies across VSAs, it results in a product that is of the same dimensionality as the component vectors from which it was derived, unlike the tensor product which has the dimensionality of its component vectors squared. When XOR is used, binding commutes: $A \otimes B = B \otimes A$.

Bundling is an addition-like operator, through which superposition of vectors is achieved. For example, vector addition followed by normalization is commonly employed as a bundling operator. Unlike binding, bundling results in a vector that is maximally similar to its component vectors. We will write the usual "+" for bundling, and the computer science "+=" for "bundle the left hand side with the right hand side and assign the outcome to the symbol on the left hand side." So for example, $S(A) +=$ $E(B)$ means "increment the semantic vector for A by the elemental vector for B using the bundling operator." This in particular is a very standard operation in training.

In the case of the spatter code, XOR is used as a binding operator. As it is its own inverse, the binding and decoding processes are identical ($\otimes = \oslash$). For bundling, the spatter code employs a majority vote: if the component vectors of the bundle have more ones than zeros in a dimension, this dimension will have a value of one, with ties broken at random (for example, bundling the vectors 011 and 010 may produce either 010 or 011). Once a vector representation for a concept has been built up by binding and/or bundling, it is possible to apply an operator that reverses the binding process to the vector as a whole.

The XOR operator used in the spatter code offers an apparent advantage over the original permutation-based implementation of PSI: both concepts and relations are represented as high-dimensional binary vectors. This suggests relatively simple ways to

Table 1. Comparison between real vector and binary vector implementation of PSI

Implementation	Real/Permutation-based	Binary
Semantic vectors $S(X)$	Real vectors ($d = 500$)	Binary vectors ($d = 16,000$)
Elemental vectors $E(X)$	Sparse ternary	Dense binary
Represent predicate R	Assign permutation P_R	Assign elemental vector $E(R)$
Reversed predicates R_{inv}	Use natural inverse P_R^{-1}	Assign new elemental vector $E(R_{inv})$
Encoding / training of relationship X R Y	$S(X) \mathrel{+}= P_R(E(Y))$ $S(Y) \mathrel{+}= P_R^{-1}(E(X))$	$S(X) \mathrel{+}= E(R) \otimes E(Y)$ $S(Y) \mathrel{+}= E(R_{inv}) \otimes E(X)$
Superposition	Vector addition	Majority vote

direct search across predicate paths of interest, such as those that have been shown useful for literature-based discovery [20]. For example, the "ISA-TREATS$_{inv}$" path, which may identify conditions treated by the class a drug belongs to, can be specified as "$S(\text{prozac}) \oslash E(\text{ISA}) \otimes E(\text{TREATS}_{inv})$." To explore the potential advantages of this formulation, we generated a binary implementation of PSI. This differs from our previous implementation in several ways, summarized in Table 1.

We are now in a position to describe our core algorithm for building the binary PSI space used in our experiments throughout the rest of this paper. The procedure is as follows:

1. **Assign an elemental vector $E(X)$ to each concept X** that occurs 100,000 times or less in the database. More frequent concepts are excluded as they tend to be uninformative, approximating use of a stop-word list. Elemental vectors are 16,000-dimensional binary vectors with a 50% chance of a one or zero in each position.
2. **Assign an elemental vector $E(R)$ to each predicate type R** excluding negations and the PROCESS_OF predicate,[1] which has shown to be uninformative. In most cases, two vectors are assigned, one for each direction of the predicate R and R_{inv}, to distinguish between the roles of the concepts involved. For a small number of symmetric predicate types, such as COEXISTS_WITH, only one vector is assigned. Note that this process differs from the original implementation using permutations as operations, since each permutation P has a natural distinct inverse P_{-1}. This is not the case for the current implementation, since XOR is its own inverse. In addition we assign a vector "GA" to represent general association.
3. **Assign a semantic vector to each concept** occurring 100,000 or fewer times. In this implementation, semantic vectors contain 16,000 real-valued variables, initially set to zero. These keep track of votes in each dimension to facilitate bundling.
4. **Statistical weighting** is applied to accentuate the influence of infrequent terms. Inverse document frequency (idf) is calculated for concepts and predicates, and applied during encoding such that general associations are weighted according to the idf of the concept concerned, while specific (typed) relations are weighted according

---

[1] This predicate occurs in predications such as "tuberculosis PROCESS_OF patient" which would create an uninformative link between most human diseases.

to the sum of the idfs of the concept and the predicate concerned. Consequently, specific relations are weighted more heavily than general relatons.

5. **Process the predications a concept occurs in:** each time a concept occurs in a predication, add (bundle) to its semantic vector the elemental vector for the other concept in the predication bound with the elemental vector for the predicate concerned. For example, when the concept fluoxetine occurs in the predication "fluoxetine TREATS major depressive disorder (MDD)," we add to S(fluoxetine) the elemental vector for TREATS bound with the elemental vector for MDD. We also encode general association by bundling the elemental vector for MDD bound with the elemental vector for general association (GA), ensuring that two concepts relating to the same third concept will have similar vectors, even if they relate to it in different ways. In symbols, we have that S(fluoxetine) += E(TREATS) $\otimes E$(MDD) + E(GA) $\otimes E$(MDD).

The PSI space was derived from a set of 22,669,964 predications extracted from citations added to MEDLINE over the past decade by the SemRep natural language processing system [21], which extracts predications from biomedical text using domain knowledge in the Unified Medical Language System [22]. For example, the predication "fluoxetine TREATS MDD" is extracted from "patients who have been successfully treated with fluoxetine for major depression." In a recent evaluation of SemRep, Kilicoglu et al. report .75 precision and .64 recall (.69 f-score) [23].

4   Analogical Retrieval

Now that we have built our PSI space, we can use it to search for relations and analogies of concepts as described in the abstract and introduction. The process for performing this search in predication space is similar to Kanerva's XOR-based analogical mapping [2]. Consider the vectors S(fluoxetine) and E(MDD):

$$S(\text{fluoxetine}) = E(\text{MDD}) \otimes E(\text{TREATS}) + E(\text{MDD}) \otimes E(\text{GA})$$
$$S(\text{fluoxetine}) \oslash E(\text{MDD}) = E(\text{MDD}) \oslash E(\text{MDD}) \otimes E(\text{TREATS})$$
$$+ E(\text{MDD}) \oslash E(\text{MDD}) \otimes E(\text{GA})$$
$$= E(\text{TREATS}) + E(\text{GA})$$

When encoding many predications, the result will be a noisy version of this vector, which should be approximately equidistant from E(TREATS) and E(GA). Therefore we would anticipate being able to search for the treatment for schizophrenia, for example, by finding the semantic vector that is closest to the vector "S(fluoxetine) $\oslash$ E(MDD) $\otimes E$(schizophrenia)." This search approximates the single-relation analogies that occur as questions in standardized tests such as the SAT, and have been the focus of recent evaluations of distributional models that estimate relational similarity (eg. [24]). However, useful predicate paths, such as the ISA-TREATS$_{\text{inv}}$ example, often involve more than one relation. The mathematical properties of the binary PSI space suggest that a similar approach can also be used to search across two relations. Consider the following steps that occur during generation of the binary PSI space:

$$S(\text{amoxicillin}) \mathrel{+}= E(\text{antibiotics}) \otimes E(\text{ISA})$$
$$S(\text{streptococcal tonsilitis}) \mathrel{+}= E(\text{antibiotics}) \otimes E(\text{TREATS}_{\text{inv}})$$
$$S(\text{prozac}) \mathrel{+}= E(\text{fluoxetine}) \otimes E(\text{ISA})$$
$$S(\text{MDD}) \mathrel{+}= E(\text{fluoxetine}) \otimes E(\text{TREATS}_{\text{inv}})$$

Assuming for the sake of simplicity that these are the only encoding operations that have taken place, an example cue could be generated as follows:

$$S(\text{amoxicillin}) \oslash S(\text{streptococcal tonsilitis})$$
$$= E(\text{ISA}) \otimes E(\text{antibiotics}) \oslash E(\text{antibiotics}) \otimes E(\text{TREATS}_{\text{inv}})$$
$$= E(\text{ISA}) \otimes E(\text{TREATS}_{\text{inv}})$$
$$S(\text{MDD}) \oslash S(\text{amoxicillin}) \oslash S(\text{streptococcal tonsilitis})$$
$$= E(\text{fluoxetine}) \otimes E(\text{TREATS}_{\text{inv}}) \oslash E(\text{TREATS}_{\text{inv}}) \otimes E(\text{ISA})$$
$$= E(\text{fluoxetine}) \otimes E(\text{ISA})$$
$$= S(\text{prozac})$$

Table 2 illustrates analogical retrieval with single and dual predicates. For single predicates (top three examples), the cue is constructed by combining $E(\text{schizophrenia})$ with the elemental and semantic vector of a pair of concepts, using XOR. The nearest semantic vector to this composite cue is in all cases related to schizophrenia by the same relation that links the example pair: emd_57445 is an experimental treatment for schizophrenia [25], syngr1 is a gene that has been associated with it [26], and certain mannerisms are relevant to the diagnosis of schizophrenia.

In the case of dual predicates (bottom three examples), the cue is constructed by combining the semantic vector for schizophrenia with the semantic vectors for a pair of concepts, using XOR. Depression is treated by antidepressants such as prozac. Similarly, schizophrenia is treated by antipsychotic agents, such as mazapertine succinate. Blood glucose fluctuation is a side effect of diabetic treatment, as impaired work performance is a side effect of drugs treating schizophrenia. Finally, chronic confusion

Table 2. Schizophrenia-related searches, single- (top 3) and dual-predicate (bottom 3). MDD=Major Depressive Disorder. Scores indicate $1-$normalized hamming distance.

Example pair	Nearest predicate	Nearest neighboring semantic vector
$S(\text{fluoxetine}) \oslash E(\text{MDD})$	$E(\text{TREATS})$	0.56 $S(\text{emd\_57445})$
$S(\text{apolipoprotein e gene})$ $\oslash E(\text{alzheimer's disease})$	$E(\text{ASSOCIATED\_WITH})$	0.76 $S(\text{syngr1})$
$S(\text{wheezing}) \oslash E(\text{asthma})$	$E(\text{DIAGNOSES})$	0.63 $S(\text{mannerism})$
$S(\text{prozac}) \oslash S(\text{MDD})$	$E(\text{ISA}) \otimes E(\text{TREATS}_{\text{inv}})$	0.54 $S(\text{mazapertine succinate})$
$S(\text{diabetes mellitus}) \oslash$ $S(\text{blood glucose fluctuation})$	$E(\text{TREATS}_{\text{inv}}) \otimes$ $E(\text{CAUSES}_{\text{inv}})$	0.55 $S(\text{impaired job performance})$
$S(\text{chronic confusion}) \oslash$ $S(\text{alzheimer's disease})$	$E(\text{ISA}) \otimes$ $E(\text{COEXISTS\_WITH})$	0.76 $S(\text{acculturation difficulty})$

occurs in dementias such as Alzheimer's, as acculturation difficulty occurs in psychotic disorders such as schizophrenia.

4.1 Evaluation

To evaluate the single-predicate approach, we extracted a set of test predications from the database using the following procedure. Firstly, a set of candidate predicates was selected. Only predicates meeting the previously-listed constraints for inclusion in our vector space model that occurred one thousand or more times in the data set were considered, leaving a total of 37 predicate types (such as DIAGNOSES). For each of these predicates, fifty predications were randomly selected taking into account the strength of association between the example pair (e.g. S(wheezing) $\oslash$ E(asthma)) and the predicate (e.g. E(DIAGNOSES)) such that ten examples were obtained for each predicate that fell into the following ranges of association strength: 0.5211-0.6, 0.61-0.7, 0.71-0.8, 0.81-0.9, 0.91-1.0. We sampled in this manner in order to test the hypothesis that better examples would have a stronger cue-to-predicate association strength, and excluded any example pairs in which this association was less than 0.5211, a value 5SD above the median similarity between a set of 5000 random vectors. Only predicates in which ten examples in each category could be found were tested, resulting in a test set of 1400 predications, fifty per eligible predicate (n=28). For each predicate, every example was tested against every other example pair (n=49) using three approaches summarized in Table 3. 68,600 searches were conducted with each approach. In each case, the nearest semantic vector (e.g. S(mannerism)) to the composite cue vector (e.g. S(wheezing) $\oslash$ E(asthma) $\otimes$ E(schizophrenia)) was retrieved, and tested for occurrence in a predication with the object of the second pair (e.g. schizophrenia), and the same predicate as the example pair (e.g. DIAGNOSES).

To evaluate the paired-predicate approach, we selected fourteen relationship pairs representing predicate paths of interest, including our recurring ISA-TREATS$_{inv}$ example, and pairs such as INHIBITS-CAUSES$_{inv}$ that are of interest for literature-based discovery [20]. For each pair, we extracted sixty example concept pairs by first selecting for each subject (e.g. prozac) occurring in a relationship of the first type (e.g. ISA) the bridging term (e.g. fluoxetine) and object (e.g. MDD) of the second relationship (e.g. TREATS$_{inv}$) with the strongest cue-to-predicate-pair association (similarity between S(prozac) $\oslash$ S(MDD) and E(ISA) $\otimes$ E(TREATS$_{inv}$)). This constraint ensured that it was possible to obtain an adequate number of examples at each cue-to-predicate-pair threshold level. These strongly associated paths were sampled at random, such that sixty example pairs were drawn for each predicate pair, with twenty of these occurring in each of the threshold levels 0.5211-0.6, 0.61-0.7, 0.71-1.0.

Each elemental predicate vector was bound to every other predicate vector, to generate a set of 5,929 paired predicate vectors, such as E(TREATS$_{inv}$) $\otimes$ E(ISA), to use for the dual-relation equivalent of the 2-STEP procedure. This and other procedures used to generate cues for this experiment are shown in Table 3. The major difference from the single-relation approach is the use of the semantic vector for both subject and object of the example pair to generate the cue. Also, the general association step does not require binding, as we would anticipate the semantic vectors for two objects associated with the same subject being similar once constructed. Each of the example pairs

Table 3. Approaches to cue vector generation. sub_1, obj_1 = subject and object from example pair. $Obj2$ = test object. E(pred_nearest) = nearest predicate vector ((1) single-predicate) or bound predicate vectors ((2) dual-predicate) to bound example pair. GA = general association.

Method	Bound cue vector	Example
1-STEP (1)	$S(sub_1) \oslash E(obj_1) \otimes E(obj_2)$	$S(\text{fluoxetine}) \oslash E(\text{MDD})$ $\otimes E(\text{schizophrenia})$
2-STEP (1)	$E(\text{pred\_nearest}) \otimes E(obj_2)$	$E(\text{schizophrenia}) \otimes E(\text{TREATS})$
GA (1)	$E(\text{GA}) \otimes E(obj_2)$	$E(\text{GA}) \otimes E(\text{schizophrenia})$
1-STEP (2)	$S(sub_1) \oslash S(obj_1) \oslash S(obj_2)$	$S(\text{prozac}) \oslash S(\text{MDD}) \oslash S(\text{schizophrenia})$
2-STEP (2)	$E(\text{pred\_nearest}) \oslash S(obj_2)$	$E(\text{ISA}) \otimes E(\text{TREATS}_{inv})$ $\oslash S(\text{schizophrenia})$
GA (2)	$S(obj_2)$	$S(\text{schizophrenia})$

(n=60) for each predicate pair was tested with the object of every other example pair in the set (n=59), for a total of 49,560 searches per method.

Approaches to cue generation are summarized in Table 3. The generated cues are intended to be similar to the vector representation of the concept (or concepts) providing a solution to an analogical problem of the form **sub_1** is to **obj_1** as **what** is to **obj_2**? 1-STEP cue generation binds the example pair to the target object directly. The 2-STEP approach first finds the nearest predicate vector (single predicates) or bound predicate vectors (dual predicates) to the example pair, and then binds this to the target object. The store of predicate vectors here acts as a "clean-up memory" (Plate 1994 [19], pg 101), removing noise from the approximate representation of the predicate (or pair of predicates) retrieved from the example pair. Finally, as a control, we retrieve the concept that our model associates most strongly with the object when the relation type is not considered (General Association, GA). As an additional control, we repeated both experiments while searching the space of elemental vectors using the elemental vector for the test object, to provide a random baseline. As this failed to produce any correct mappings in the vast majority of cases, the results are not shown.

4.2 Results

The results of the single predicate experiment are shown in Fig. 1 (left). The y-axis shows the mean number of test cases in which the retrieved concept occurred in a predication with the test target in which the predicate matched that linking the example pair. Both the 1-STEP and 2-STEP approaches are sensitive to the strength of association between the example pair and the predicate that links them. As might be expected, an intermediate step utilizing clean-up memory improves performance in the 2-STEP approach, particularly as the cue-to-predicate association drops. These results show that an example concept pair can be used to prime search to retrieve concepts that are related to a cue concept in a particular way, with (2-STEP) or without (1-STEP) retrieving a representation of the relationship concerned. This approach is particularly effective with example pairs that have a strong association to the representation of the predicate of interest. The GA approach retrieves a correct mapping less frequently, and is not sensitive to cue-to-predicate association.

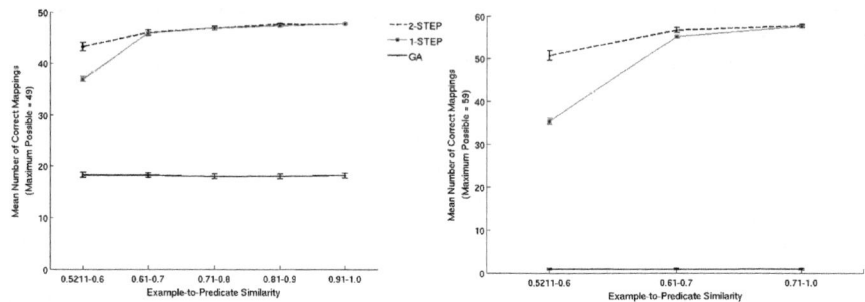

Fig. 1. Analogical retrieval: single (left) and dual (right) predicates. Error bars = standard error

Fig. 1 (right) shows the results of the dual-predicate experiment, which are similar to those for single-relation searches: at stronger cue-to-predicate associations, correct mappings are found in most cases, whereas with cue-to-predicate associations closer to those anticipated between randomly generated vectors, performance falls. This drop in performance is mitigated to some extent by the use of the 2-STEP approach, in which clean-up memory is used to obtain the original vector representation of the paired relationship concerned. The GA approach is less effective here. While these results do indicate search-by-example is effective in certain cases, the constraint that cue-to-predicate strength should fall in the upper strata limits this approach to a small set of example cues. For example, in the case of the ISA-TREATS$_{inv}$ predicate pair, the distribution of cue-to-predicate associations in the set (n=114,658) from which our example cues were sampled (which itself included only the best example for each subject) skews leftward, with a median association strength of 0.522. A similar distribution was observed for single-predicate cues. It is possible to compensate for this using the 2-STEP approach, but this is not ideal for paired relations: with r relations the 2-STEP approach requires searching through r^2 possible predicate pairs. However, as each weak example should have some association with the desired path, we would anticipate the superposition of several weak examples generating a vector with a stronger cue-to-predication-path strength than any of its components. To evaluate this hypothesis, we generated a second set of example pairs for the ISA-TREATS$_{inv}$ predicate path. These examples were drawn from the aforementioned set, with the inclusion criterion that their cue-to-predicate association must fall in the weakest category (0.5211 - 0.6). For each example, we measured the cue-predicate association of the example pair ($S(\text{sub}_1) \oslash S(\text{obj}_1)$). As we added new examples, we also measured the association strength between the superposition of all examples up to this point ($S(\text{sub}_1) \oslash S(\text{obj}_1) + \ldots + S(\text{sub}_n) \oslash S(\text{obj}_n)$) and the desired predicate ($E(\text{ISA}) \otimes E(\text{TREATS}_{inv})$).

The results of this experiment are shown in Fig 2 (left), which shows a rapid rise in cue-to-predicate strength (solid line) as weak examples are added to the superposition. The strength of this association quickly exceeds the cumulative mean (dashed line) association strength of all of the examples added up to that point (individual dots). As shown in Fig. 2 (right), this effect is also observed with respect to performance on the ISA-TREATS$_{inv}$ test examples (n=60). This is a particularly important result from the "generalized quantum" point of view. We have used repeated binding and bundling to

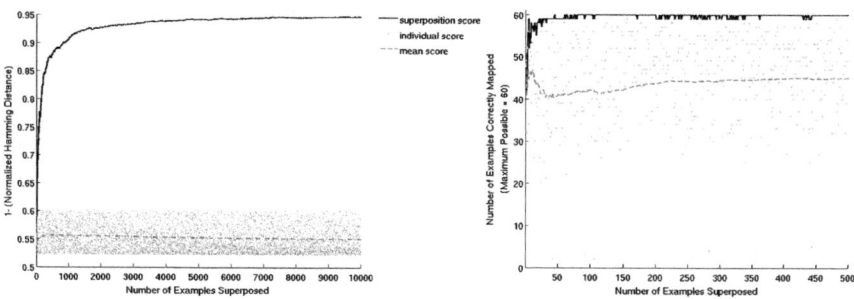

Fig. 2. Superposition: cue-predicate association (left), correct mappings (right)

create a superposition of compound systems that has not been (and probably cannot be) represented as a product of two individual simple systems. In the quantum literature, this phenomenon is known as "entanglement". Thus our experiments demonstrate that several weak example relationships can be superimposed to obtain an entangled representation of the typed relation which is a much more accurate guide for inferring new examples.

5 Discussion

In this paper, we show that relational similarity emerges as a natural consequence of the PSI approach. This similarity is sufficient to solve proportional analogy problems stretching across one and two relations, given either a strong example with well-preserved similarity to the relation(s) of interest, or a set of weaker examples. These findings are pertinent to our ongoing research in the area of literature-based discovery and abductive reasoning. Previously, we have discussed various forms of abductive reasoning and constraints operative in such reasoning, and proposed that similarity of some kind is often of importance in finding a link between a starting point of an inquiry and fruitful novel connection to the starting point [27]. The associations are usually weak and indirect, but likely critical in making the connection. Analogy is one form of such indirect connection. An analogy and the starting point have relationships in common [28] so presumably finding cases of common relations is at the heart of analogy retrieval. There have been several implementations of vector encoding to accomplish analogical reasoning [29], [30]. These modeling efforts aim to address several aspects of analogical reasoning: retrieving potential analogies, mapping the elements of the potential target analogy to the elements of the starting point, and making inferences about the starting point from the target analogy. Our goals are more modest in some respects and more ambitious in others. We are initially only concerned with retrieving potential analogies, but we aim to do this on a large scale using large numbers of predications that have been automatically extracted from the biomedical literature, while most of the models of analogies have worked with small sets of custom-constructed predications relating to a few stories. Through analogical retrieval, we are able to direct search across predicate paths that have been shown to be useful for literature-based discovery [20], without incurring an exponential increase in the size of the search space when more than one relationship is considered. The facility for search of this nature is an emergent property of the PSI model: candidates for retrieval are identified on the basis of their similarity to a vector representing a novel relation type, composed from elemental relations during the process of model generation. An approximation of this vector is inferred from the superposition of a set of example pairs, providing an efficient and accurate mechanism for directed search.

6   Conclusion

In this paper, we show that accurate example-based analogical retrieval across single and dual-predicate paths emerges as a natural consequence of the encoding of typed relations in high-dimensional vector space. Given a suitable example pair, or set of less suitable example pairs, it is possible to retrieve with accuracy concepts that relate to another concept in the same way as the concepts in the example pair relate to one another, even if this relationship involves two relations and a third bridging concept. In the case of dual relations, search is achieved without the need to retrieve either the bridging concept or the relations involved. The size of the search space does not increase when dual-relation paths are sought, providing an efficient means to direct predication-based search toward pathways of interest for literature-based discovery.

Acknowledgements. This research was supported in part by the US National Library of Medicine grant (R21LM010826-01). The authors would also like to acknowledge Google, Inc. for their support of author DW's ongoing research on the subject.

References

1. Cohen, T., Widdows, D., Schvaneveldt, R.W., Rindflesch, T.C.: Logical leaps and quantum connectives: Forging paths through predication space. In: AAAI-Fall 2010 Symposium on Quantum Informatics for Cognitive, Social, and Semantic Processes, pp. 11–13 (November 2010)
2. Kanerva, P.: Hyperdimensional computing: An introduction to computing in distributed representation with high-dimensional random vectors. Cognitive Computation 1(2), 139–159 (2009)
3. Van Rijsbergen, C.J.: The Geometry of Information Retrieval. Cambridge University Press, Cambridge (2004)
4. De Vine, L., Bruza, P.: Semantic Oscillations: Encoding Context and Structure in Complex Valued Holographic Vectors. In: Quantum Informatics for Cognitive, Social, and Semantic Processes, QI 2010 (2010)
5. Cohen, T., Schvaneveldt, R., Widdows, D.: Reflective Random Indexing and indirect inference: A scalable method for discovery of implicit connections. Journal of Biomedical Informatics 43(2), 240–256 (2010)
6. Landauer, T.K., Dumais, S.T.: A solution to Plato's problem: The latent semantic analysis theory of acquisition, induction, and representation of knowledge. Psychological Review 104, 211–240 (1997)
7. Swanson, D.R.: Two Medical Literatures that are Logically but not Bibliographically Connected. Prog. Lipid. Res. 21, 82 (2007)
8. DiGiacomo, R.A., Kremer, J.M., Shah, D.M.: Fish-oil dietary supplementation in patients with Raynaud's phenomenon: a double-blind, controlled, prospective study. The American Journal of Medicine 86, 158–164 (1989)
9. Peirce, C.S.: Abduction and Induction. In: Buchler, J. (ed.) Philosophical Writings of Peirce. Routledge, New York (1940)

10. Swanson, D.R., Smalheiser, N.R.: An interactive system for finding complementary literatures: a stimulus to scientific discovery. Artificial Intelligence 91, 183–203 (1997)
11. Weeber, M., Kors, J.A., Mons, B.: Online tools to support literature-based discovery in the life sciences. Briefings in Bioinformatics 6(3), 277–286 (2005)
12. Kanerva, P., Kristofersson, J., Holst, A.: Random indexing of text samples for latent semantic analysis. In: Proceedings of the 22nd Annual Conference of the Cognitive Science Society, vol. 1036 (2000)
13. Gordon, M.D., Dumais, S.: Using latent semantic indexing for literature based discovery. JASIS 49, 674–685 (1998)
14. Bruza, P., Cole, R., Song, D., Bari, Z.: Towards Operational Abduction from a Cognitive Perspective, vol. 14. Oxford Univ. Press, Oxford (2006)
15. Cohen, T., Schvaneveldt, R., Rindflesch, T.: Predication-based Semantic Indexing: Permutations as a Means to Encode Predications in Semantic Space. In: Proceedings of the AMIA Annual Symposium, San Francisco (2009)
16. Kanerva, P.: Binary spatter-coding of ordered K-tuples. In: Vorbrüggen, J.C., von Seelen, W., Sendhoff, B. (eds.) ICANN 1996. LNCS, vol. 1112, pp. 869–873. Springer, Heidelberg (1996)
17. Gayler, R.W.: Vector Symbolic Architectures answer Jackendoff's challenges for cognitive neuroscience. In: Slezak, P. (ed.) ICCS/ASCS International Conference on Cognitive Science, 133-138 (2003)
18. Smolensky, P.: Tensor product variable binding and the representation of symbolic structures in connectionist systems. Artificial Intelligence 46(1), 159–216 (1990)
19. Plate, T.A.: Holographic Reduced Representation: Distributed Representation for Cognitive Structures. CSLI Publications, Stanford (2003)
20. Hristovski, D., Friedman, C., Rindflesch, T.C., Peterlin, B.: Exploiting semantic relations for literature-based discovery. In: AMIA Annual Symposium Proceedings, pp. 349–353 (2006)
21. Rindflesch, T.C., Fiszman, M.: The interaction of domain knowledge and linguistic structure in natural language processing: interpreting hypernymic propositions in biomedical text. Journal of Biomedical Informatics 36, 462–477 (2003)
22. Bodenreider, O.: The unified medical language system (UMLS): integrating biomedical terminology. Nucleic Acids Research 32, D267 (2004)
23. Kilicoglu, H., Fiszman, M., Rosemblat, G., Marimpietri, S., Rindflesch, T.C.: Arguments of nominals in semantic interpretation of biomedical text. In: Proceedings of the 2010 Workshop on Biomedical Natural Language Processing, pp. 46–54 (2010)
24. Turney, P.D.: Measuring semantic similarity by latent relational analysis. In: Proceedings of the Nineteenth International Joint Conference on Artificial Intelligence (IJCAI 2005), Edinburgh, Scotland, pp. 1136–1141 (2005)
25. Huber, M.T., Gotthardt, U., Schreiber, W., Krieg, J.C.: Efficacy and safety of the sigma receptor ligand EMD 57445 (panamesine) in patients with schizophrenia: an open clinical trial. Pharmacopsychiatry 32(2), 68–72 (1999)
26. Verma, R., Kubendran, S., Das, S.K., Jain, S., Brahmachari, S.K.: SYNGR1 is associated with schizophrenia and bipolar disorder in southern India. Journal of Human Genetics 50(12), 635–640 (2005)
27. Schvaneveldt, R., Cohen, T.: Abductive Reasoning and Similarity. In: Ifenthaler, D., Seel, N.M. (eds.) Computer Based Diagnostics and Systematic Analysis of Knowledge, Springer, New York (2010)
28. Gentner, D.: Structure-mapping: A theoretical framework for analogy. Cognitive Science 7, 155–170 (1983)
29. Plate, T.A.: Analogy retrieval and processing with distributed vector representations. Expert Systems 17(1), 29–40 (2000)
30. Eliasmith, C., Thagard, P.: Integrating structure and meaning: A distributed model of analogical mapping. Cognitive Science 25(2), 245–286 (2001)

Spectral Composition of Semantic Spaces

Peter Wittek and Sándor Darányi

Swedish School of Library and Information Science
Göteborg University & University of Borås
Allégatan 1, 50190 Borås, Sweden
peterwittek@acm.org, sandor.daranyi@hb.se

Abstract. Spectral theory in mathematics is key to the success of as diverse application domains as quantum mechanics and latent semantic indexing, both relying on eigenvalue decomposition for the localization of their respective entities in observation space. This points at some implicit "energy" inherent in semantics and in need of quantification. We show how the structure of atomic emission spectra, and meaning in concept space, go back to the same compositional principle, plus propose a tentative solution for the computation of term, document and collection "energy" content.

1 Introduction

In quantum mechanics (QM), the spectrum is the set of possible outcomes when one measures the total energy of a system. Solutions to the time-independent Schrödinger wave equation are used to calculate the energy levels and other properties of particles. A non-zero solution of the wave equation is called an eigenenergy state, or simply an eigenstate. The set of eigenvalues $\{E_j\}$ is called the energy spectrum of the particle. This energy spectrum can be mapped to frequencies in the electromagnetic spectrum.

In this paper, we argue that by decomposing a semantic space, one can gain a "semantic spectrum" for each term that makes up the space. This makes sense for the following reason: mapping spectra to the electromagnetic spectrum is a unification effort to match energy and intellectual input stored in documents by modelling semantics on QM. Energy is a metaphor here, lent from machine learning which imitates pattern recognition and pattern naming in cognitive space. We adopted this as our working hypothesis based on [1].

To this end, we ascribe significance to two aspects of the above parallel. Both make the comparison between semantics and QM reasonable. The first is an alleged similarity between them, namely eigendecomposition and related methods leading to meaningful conclusions in both. The second is the evolving nature of QM and semantic systems, based on interactions among constituents, leading to structuration. The insights we offer in this paper do not rely on extensive quantitative benchmarks. Instead, the paper reports our initial foray into exploring the above metaphor.

This paper is organized as follows. Section 2 discusses core concepts in QM relevant to this treatise. Section 3 gives an overview of semantic spaces in general

D. Song et al. (Eds.): QI 2011, LNCS 7052, pp. 60–70, 2011.

and Section 4 describes their spectral composition in particular, including their treatment as observables, corpus and term semantic spectra, and indications for future work such as evolving semantics. Section 5 sums up the conclusions.

2   Related Concepts in Quantum Mechanics and Spectroscopy

In quantum mechanics, observables are not necessarily bounded, self-adjoint operators and their spectra are the possible outcomes of measurements. The Schrödinger wave equation is an equation that describes how the quantum state of a physical system changes over time. Approximate solutions to the time-independent Schrödinger wave equation are commonly used to calculate the energy levels and other properties of atoms and molecules. From this, the emission spectrum is easy to calculate.

Emission is the process by which two quantum mechanical states of a particle become coupled to each other through a photon, resulting in the production of light. The frequency of light emitted is a function of how far away in energy the two states of the system were from each other, so that energy is conserved: the energy difference between the two states equals the energy carried off by the photon (Figure 1).

Since the emission spectrum is different for every element of the periodic table, it can be used to determine the composition of a material. In general, spectroscopy is the study of the interaction between matter and radiated energy. A subset of spectroscopic methods, called spectrophotometry, deals with visible light, near-ultraviolet, and near-infrared wavelengths. For the rest of this paper, we limit ourselves to visible spectroscopy, because this approach focuses on the electronic orbitals (i.e., where the electrons can be found), whereas, for instance, infra-red spectroscopy is concerned with the internal motions of the molecule (how the bonds stretch, angles bend, etc.).

A spectrogram is a spectral representation of an electromagnetic signal that shows the spectral density of the signal. An example is astronomical spectroscopy that studies the radiation from stars and other celestial objects (Figure 2). While discrete emission bands do not show clearly, the intensity of certain wavelengths indicates the composition of the observed object. The emission lines are caused by a transition between quantized energy states and theoretically they look very sharp, they do have a finite width, i.e. they are composed of more than one wavelength of light. This spectral line broadening has many different causes, with the continuum of energy levels called "spectral bands". The bands may

Fig. 1. The emission spectrum of hydrogen

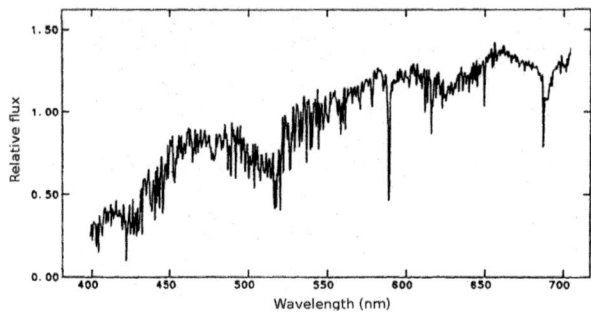

Fig. 2. The visible spectrogram of the red dwarf EQ Vir (figure adapted from [2])

overlap. Band spectra are the combinations of many different spectral lines, resulting from rotational, vibrational and electronic transitions.

3   A Brief Overview of Semantic Spaces

We regard semantic spaces as algebraic models for representing terms as vectors. The models capture term semantics by a range of mathematical relations and operations. Language technology makes extensive use of semantic spaces. Among the reasons are the following:

- The semantic space methodology makes semantics computable allowing a definition of semantic similarity in mathematical terms. Sparsity plays a key role in most semantic spaces. A term-document vector space (see below), for instance, is extremely sparse and therefore it is a feasible option for large-scale collections.
- Semantic space models also constitute an entirely descriptive approach to semantic modelling relying on the distributional hypothesis. Previous linguistic or semantic knowledge is not required.
- The geometric metaphor of meaning inherent in a vector space kind of model is intuitively plausible, and is consistent with empirical results from psychological studies. This relates especially to latent semantic indexing (see below) [3]. A link has also been established to Cognitive Science [4].

While there are several semantic space models, we restrict our discussion to the following two major kinds: term-document vector spaces [5] and latent semantic indexing (LSI, [6]); and the hyperspace analogue to language (HAL, [7]).

The coordinates in the vector of a term in a term-document space record the number of occurrences of the term in the document assigned to that particular dimension. Instead of plain term frequencies, more subtle weighting schemes can be applied, depending on the purpose. The result is an $m \times n$ matrix A, where m is the number of terms, and n is the number of documents. This matrix is extremely sparse, with only $1 - 5\%$ of the entries being non-zero. This helps scalability, but has an adverse impact on modelling semantics. For instance, in measuring similarity with a cosine function between the term vectors, we often end up with a value of zero, because the vectors do not co-occur in any of the documents of the collection, although they are otherwise related. To overcome this problem, LSI applies dimension reduction by singular value decomposition (SVD). The term-document matrix A can be decomposed as $A = U\Sigma V^{\mathrm{T}}$, where U is an $m \times m$ unitary matrix, Σ is an $m \times n$ diagonal matrix with nonnegative real numbers, the singular values, on the diagonal, and V is an $n \times n$ unitary matrix. By truncating the diagonal of Σ, keeping only the k largest singular values, we get the rank-k approximation of A, $A_k = U_k\Sigma_k V_k^{\mathrm{T}}$. This new space, while not sparse, reflects semantic relations better [3]. Apart from LSI, a term co-occurrence matrix is another alternative to overcome the problem of sparsity. It is obtained by multiplying A with its own transpose, A^{T}.

The HAL model considers context only as the terms that immediately surround a given term. HAL computes an $m \times m$ matrix H, where m is the number of terms, using a fixed-width context window that moves incrementally through a corpus of text by one word increment ignoring punctuation, sentence and paragraph boundaries. All terms within the window are considered as co-occurring with the last word in the window with a strength inversely proportional to the distance between the words. Each row i in the matrix represents accumulated weights of term i with respect to other terms which preceded i in a context window. Similarly, column i represents accumulated weights with terms that appeared after i in a window. Dimension reduction may also be performed on this matrix.

We note in passing that there exists a little recognized constraint of the model in testing: for a match between theories of word semantics and semantic spaces, a semantic space is a statistical model of word meaning observed [8]. For its workings, it has to match a reasonably complex theory of semantics; but whereas Lyons regarded meaning a composite [9], i.e. a many-faceted complex phenomenon, the distributional hypothesis [10] as the sole semantic underpinning of eigenmodels is anything but complex and must be hence deficient. One can use it as long as there is nothing else available but, at the same time, one must not stop looking for a more comprehensive model. It holds in this sense that we look at the validity and some consequences of the semantic collapse model based on quantum collapse, treating semantic deep structure as an eigenvalue spectrum.

4   Spectral Composition of Semantic Spaces

4.1   Semantic Spaces as Observables

Our line of thought is as follows: in QM, atoms have ground states low on energy, and excited states high on it. Such states are expressed as separate spectral (latent) structures, based on the way they can be identified. By analogy a term should have a "ground state" and may have several "excited states" as well, all in terms of spectra.

In what follows, we regard a semantic space an observable. This being a real or a complex space, its spectrum will be the set of eigenvalues. If we decompose a semantic space we get the so-called concept space or topic model in which terms map to different locations due to their different composition. We identify this latent topic mixture in LSI with the energy eigenstructure in QM. This means that more prevalent hidden topics correspond to higher energy states of atoms and molecules.

Identifying "excited states" of word forms with homonyms, and word sense disambiguation with observation, the above shows resemblance with the quantum collapse of meaning described by [8]. They argue that a sense can be represented as a density matrix which is quite easily derived from summing the HAL matrices of the associated contexts. In addition, a probability can be ascribed the to a given sense. For example, the density matrix ρ for the meaning of a word can be formalized at the following linear combination:$\rho = p_1\rho_1 + \ldots + p_m\rho_m$, where each i is a basis state representing one of the m senses of the term and the probabilities p_i sum to unity. This is fully in accord with QM whereby a density matrix can be expressed as a weighted combination of density matrices corresponding to basis states. Context is modelled as a projection operator which is applied to a given density matrix corresponding to the state of a word meaning resulting in its 'collapse'. The probability of collapse p is a function of the scalar quantity resulting from matching. The analogy with orthodox QM is the following - a projection operator models a measurement on a quantum particle resulting in a collapse onto a basis state. Spectral decomposition by SVD also allows the description of a word as the sum of eigenstates using the bra-ket terminology [11]. The formal description is similar to the above. Projection operators are defined by singular vectors. These are orthogonal.

The semantic space must be Hermitian to pursue the metaphor of an observable in a quantum system. The sum of a HAL space H and its transpose is a Hermitian matrix [11]. A different approach is to pad the corresponding matrix of a term-document space A with zeros to make an operator map a Hilbert space onto itself, and then use a product with its own transpose as the Hermitian operator [12]. For the rest of the paper, we adopt a similar approach, taking the term co-occurrence matrix AA^{T}, which is a Hermitian operator. For symmetric and Hermitian matrices, the eigenvalues and singular values are obviously closely related. A nonnegative eigenvalue, $\lambda \geq 0$, is also a singular value, $\sigma = \lambda$. The corresponding vectors are equal to each other, $u = v = x$. A negative eigenvalue, $\lambda < 0$, must reverse its sign to become a singular value, $\sigma = |\lambda|$. One of the corresponding singular vectors is the negative of the other, $u = -v = x$. Hence a singular value decomposition and an eigendecomposition coincide.

4.2   Semantic Spectrum

In a metaphoric sense, words in an eigendecomposition are similar to chemical compounds: as both are composed of doses of latent constituents, the dosimetric view applies to them. The field that studies substances of unknown composition is called spectrometry. By analyzing their spectra, components of such substances can be identified because each chemical component has a unique "fingerprint".

The case of a semantic spectrum is not unlike. We performed the eigendecomposition of the term co-occurrence matrix of the Reuters-21578 collection. There are many other methods to capture the latent constituents of terms, for instance random indexing [13], latent Dirichlet allocation [14], or spherical k-means [15]. It is an open question which method captures the latent structure best. We use eigendecomposition due to its similarity to spectrometry. The term co-occurrence matrix is a Hermitian operator, hence the eigenvalues are all real-valued. Since the term co-occurrence matrix does not have an underlying physical meaning, we mapped the eigenvalues to the visible spectrum. If 400nm is the lowest visible wavelength and 700nm is the highest, then, assuming that the lowest eigenvalue is approximately zero, and λ_{max} denotes the highest eigenvalue, the mapping is performed by $F(x) = 400 + x \frac{700-400}{\lambda_{max}}$. The resulting spectrum is plotted in Figure 3(a). By this mapping one obtains a visual snapshot of an unknown topic composition.

In other words, by this metaphor we regarded the semantic spectrum of the above test collection as a composite, a sum of spectra of elementary components, which would correspond to individual elements in a chemical compound in spectrophotometry. This representation stresses the similarity of chemical composition of elements to the semantic composition of words.

We propose matching spectral components to terms based on their proximity to latent variables. This creates individual, albeit overlapping, spectra for every term. Having used a 0.05 threshold value of the cosine dissimilarity measure between term vectors and eigenvectors, if the cosine was above this value, we

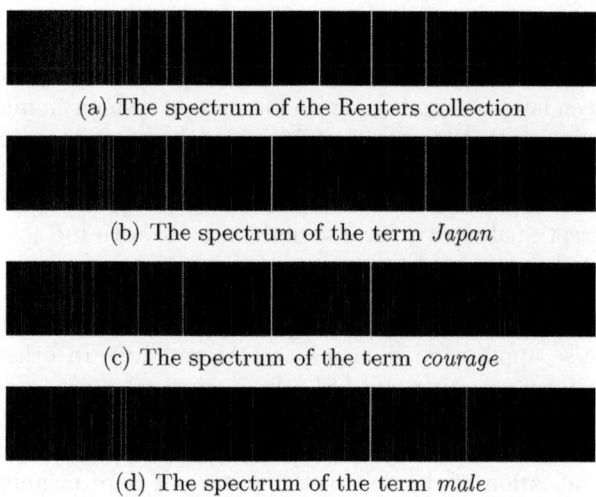

(a) The spectrum of the Reuters collection

(b) The spectrum of the term *Japan*

(c) The spectrum of the term *courage*

(d) The spectrum of the term *male*

Fig. 3. The spectrum of the collection and of different words. Higher energy states correspond to the right end of the spectrum.

added the corresponding scaled eigenvalue to the term's spectrum. In this regard, term spectra may overlap, and their simple sum will provide the spectrum of the collection. This metaphor does not account for more complex chemical bonds that create the continuous bands as pictured in Figure 2.

By such experimentation, one can end up with interesting interpretation problems. For instance, the term *Japan* (Figure 3(b)) has a high wavelength component, and a number of low wavelengths. This means that by the formula $E_{\text{photon}} = h\nu$, where h is Planck's constant and ν is the frequency (the inverse of wavelength multiplied by the speed of light), the term has one low-energy state which it is likely to take, and a number of other, high-energy states which it takes given an appropriate context. In its low-energy states the term is likely to refer to the country itself, whereas the less frequently encountered contexts may activate one of the four nominal and one verbal senses listed in WordNet. In other words, the term was correctly treated as a homonym by considering its senses as atoms in a molecule.

Another example, the term *courage* does not have a true low-energy state, it takes only higher-energy configurations. Here our tentative suggestion is that eigendecompositon does not distinguish between molecular or atomic electron orbits, hence future research may indicate that such high energy states are typical for terms treated as atoms (Figure 3(c)).

The term *male* can take two fairly low-energy states, but very few higher ones (Figure 3(d)). Since this word has three nominal and three verbal senses in WordNet, it is a reasonable working hypothesis to say that the term was treated as a molecule with six states. We trust that by more experimentation, we will gain better insight into the art of semantic spectrogram interpretation.

4.3   Evolving Semantics and Considerations for Future Work

A related aspect of our approach is the quest to formalize corpus dynamics, in line with the recommendations spelled out by [16], also keeping the possible differences between language and quantum interaction systems in mind. We depart from the assumption that two types of dynamics characterize any text document collection: external forces leading to its expansion, and the inherent quality in terms and their agglomerates called their meaning. We offer two observations why this inherent quality may have something to do with the concept of energy (a.k.a. work content):

- Interestingly, spectral theory in mathematics has been key to the success of as diverse application domains as QM and LSI. In other words, both the Schrodinger equation and LSI rely on eigenvalue decomposition for the localization of their respective entities in observation space. This points at some implicit "energy" inherent in semantics and in need of quantification. Another indication of the "energetic" nature of word meaning comes from dynamic semantics where it is regarded as an agent or promoter of change [17,18]. However, contextual and referential theories of word meaning [10,19] currently used in applications trying to capture and exploit semantic content focus on the *quantities* of qualities only, and may therefore miss part of the underlying framework;

- The phenomenon of language change and its modelling [20] necessitates a coherent explanation of the dynamics of evolving collections. In line with the above, since any matrix has an eigendecomposition and therefore a latent structure, evolving vector spaces of terms and documents follow directly from variable matrix spectra. However, this has implications for modelling semantics on QM, plus offers an illustration to the problem of assigning an "energetic" nature to word meaning. Namely, whereas Salton's dynamic library model [21], except for mass, already embodied all the key concepts of Newtonian mechanics, it is exactly this missing element which prevents one from constructing time-dependent term and document potential fields, and hence evolving "energy" landscapes. Also, without assuming that terms and documents have specific "masses" and corresponding "energies", it is very difficult to explain how intellectual work can be stored in documents and collections. In other words, unless one comes up with a better solution to the problem of how thinking amounts to work, one must assume that work as the line integral of force needs a language model which utilizes the concepts of distance, velocity, acceleration, mass, force and potential.

The implication is that if we want to be coherent, applying QM for a better understanding of meaning begs for the concept of a term-specific mass. However, such specific values cannot be extracted from an evolving environment, therefore they must reside somewhere else, e.g. in a stable environs such as an ontology, from where they can "charge" entities as their forms with content. This would amount to a challenge to the current view on semantic spaces which strives to explain the presence of all the meaning in vector spaces by term context only, and would resemble a referential model of word semantics instead. A series of semantic spectrograms, i.e. snapshots taken of collection content over time could display this evolving latent "energy" structure, and illustrate our point. In such an environment, term "energies" cannot be either constant or specific though, a contradiction to be explored.

In QM, it is the Hamiltonian which typically describes the energy stored in a system. With the above caveat, it is evident that in order to experiment with the dynamic aspect of meaning, one needs to take a look at the Hamiltonian of a collection. Further because in the above experiment, we identified the superposition of term states in the absence of an observer with that of homonyms in need of disambiguation, the same word form with different senses invites the parallel of molecular orbitals, and hence the use of the molecular Hamiltonian. This is the equation representing the energy of the electrons and nuclei in a molecule, a Hermitian operator which, together with its associated Schrödinger equation, plays a central role in computational chemistry and physics for computing properties of molecules and their aggregates.

At the same time it is necessary to point out that, whereas the demonstrated applicability of QM to semantic spaces implies the presence of some force such as lexical attraction [22] or anticipated term mass [23], because of the "energetic" explanation we can calculate with two kinds of attraction between terms only, i.e. one caused by polarity and leading to the Coulomb potential, the other caused by mass and leading to gravitational potential. But whereas there is hope that some aspect of vocabularies can be associated in the future with the role mass plays in physics, we do not know of any attempts to explain vector spaces in terms of polarity such as negative and positive electric charges unless one considers absence and presence in a binary matrix as such. However, then some kind of existential polarity is modelled by the wrong numerical kit, but nevertheless, as the results prove, the metaphor works: the expression could be constructed. Meanwhile, semantics modelled on QM also works, but we do not know why, as according to our current understanding, with this many ill fits between physics and language, it should not. These contradictions call for continued research.

5   Conclusions

Apart from semantic spectrograms bringing closer the idea of mathematical energy, a frequent concept in machine learning and structured prediction [1], our approach has the following attractive implications with their own research potential:

- Studying and eventually composing semantic functions from matrix spectra is a new knowledge area where the mathematical objects used, i.e. functions, have a higher representation capacity than vectors. This surplus can be used for the encoding of different aspects of word and sentence semantics not available by vector representation, and in general opens up new possibilities for knowledge representation;
- This form of semantic content representation provides new opportunities for optical computing, including computation by colours [24];
- Connecting QM and language by the concept of energy, represented in the visual spectrum, has a certain flair which goes beyond the paedagogical usefulness of the metaphor. Namely, considering semantics as a kind of energy and expressing it explicitly as such brings the very idea of intellectual work stored in documents one step closer to measurable reality, of course with all the foreseeable complications such an endeavour might entail.

Acknowledgement. This work was partially funded by Amazon Web Services and the large-scale integrating project Sustaining Heritage Access through Multivalent ArchiviNg (SHAMAN) which is co-funded by the European Union (Grant Agreement No. ICT-216736).

References

1. LeCun, Y., Chopra, S., Hadsell, R.: A tutorial on energy-based learning. In: Predicting Structured Data, pp. 1–59. MIT Press, Cambridge (2006)
2. Pettersen, B., Hawley, S.: A spectroscopic survey of red dwarf flare stars. Astronomy and Astrophysics 217, 187–200 (1989)

3. Landauer, T., Dumais, S.: A solution to Plato's problem: The latent semantic analysis theory of acquisition, induction, and representation of knowledge. Psychological Review 104(2), 211–240 (1997)
4. Gärdenfors, P.: Conceptual spaces: The geometry of thought. The MIT Press, Cambridge (2000)
5. Salton, G., Wong, A., Yang, C.: A vector space model for information retrieval. Journal of the American Society for Information Science 18(11), 613–620 (1975)
6. Deerwester, S., Dumais, S., Furnas, G., Landauer, T., Harshman, R.: Indexing by latent semantic analysis. Journal of the American Society for Information Science 41(6), 391–407 (1990)
7. Lund, K., Burgess, C.: Producing high-dimensional semantic spaces from lexical co-occurrence. Behavior Research Methods Instruments and Computers 28, 203–208 (1996)
8. Bruza, P., Woods, J.: Quantum collapse in semantic space: interpreting natural language argumentation. In: Proceedings of QI 2008, 2nd International Symposium on Quantum Interaction. College Publications, Oxford (2008)
9. Lyons, J.: Introduction to theoretical linguistics. Cambridge University Press, New York (1968)
10. Harris, Z.: Distributional structure. In: Harris, Z. (ed.) Papers in Structural and Transformational Linguistics. Formal Linguistics, pp. 775–794. Humanities Press, New York (1970)
11. Bruza, P., Cole, R.: Quantum logic of semantic space: An exploratory investigation of context effects in practical reasoning. In: Artemov, S., Barringer, H., d' Avila Garcez, A.S., Lamb, L., Woods, J. (eds.) We Will Show Them: Essays in Honour of Dov Gabbay. College Publications (2005)
12. Aerts, D., Czachor, M.: Quantum aspects of semantic analysis and symbolic artificial intelligence. Journal of Physics A: Mathematical and General 37, L123–L132 (2004)
13. Kanerva, P., Kristofersson, J., Holst, A.: Random indexing of text samples for latent semantic analysis. In: Proceedings of CogSci 2000, 22nd Annual Conference of the Cognitive Science Society, Philadelphia, PA, USA, vol. 1036 (2000)
14. Blei, D., Ng, A., Jordan, M.: Latent Dirichlet allocation. The Journal of Machine Learning Research 3, 993–1022 (2003)
15. Dhillon, I., Modha, D.: Concept decompositions for large sparse text data using clustering. Machine Learning 42(1), 143–175 (2001)
16. Kitto, K., Bruza, P., Sitbon, L.: Generalising unitary time evolution. In: Bruza, P., Sofge, D., Lawless, W., van Rijsbergen, K., Klusch, M. (eds.) QI 2009. LNCS, vol. 5494, pp. 17–28. Springer, Heidelberg (2009)
17. Beaver, D.: Presupposition and assertion in dynamic semantics. CSLI publications, Stanford (2001)
18. van Eijck, J., Visser, A.: Dynamic semantics. In: Zalta, E.N. (ed.) The Stanford Encyclopedia of Philosophy (2010)
19. Frege, G.: Sense and reference. The Philosophical Review 57(3), 209–230 (1948)
20. Baker, A.: Computational approaches to the study of language change. Language and Linguistics Compass 2(3), 289–307 (2008)
21. Salton, G.: Dynamic information and library processing (1975)
22. Beeferman, D., Berger, A., Lafferty, J.: A model of lexical attraction and repulsion. In: Proceedings of ACL 1997, 35th Annual Meeting of the Association for Computational Linguistics, Madrid, Spain, pp. 373–380. ACL, Morristown (1997)

23. Shi, S., Wen, J., Yu, Q., Song, R., Ma, W.: Gravitation-based model for information retrieval. In: Proceedings of SIGIR 2005, 28th International Conference on Research and Development in Information Retrieval, Salvador, Brazil, pp. 488–495. ACM, New York (2005)

24. Dorrer, C., Londero, P., Anderson, M., Wallentowitz, S., Walmsley, I.: Computing with interference: all-optical single-query 50-element database search. In: Proceedings of QELS 2001, Quantum Electronics and Laser Science Conference, pp. 149–150 (2001)

Dynamic Optimization with Type Indeterminate Decision-Maker: A Theory of Multiple-self Management

Ariane Lambert-Mogiliansky[1] and Jerome Busemeyer[2]

[1] Paris School of Economics
alambert@pse.ens.fr
[2] Indiana University
jbusemey@indiana.edu

Abstract. We study the implications of quantum type indeterminacy for a single agent's dynamic decision problem. When the agent is aware that his decision today affects the preferences that will be relevant for his decisions tomorow, the dynamic optimization problem translates into a game with multiple selves and provides a suitable framework to address issues of self-control.. The TI-model delivers a theory of self-management in terms of decentralized Bayes-Nash equilibrium among the potential eigentypes(selves). In a numerical example we show how the predictions of the TI-model differ from that of a classical model. In the TI-model choices immediately (without additional structure) reflect self-management concerns. In particular, what may be perceived as a feature of dynamic inconsistency, may instead reflect rational optimization by a type indeterminate agent.

"The idea of self-control is paradoxical unless it is assumed that the psyche contains more than one energy system, and that these energy systems have some degree of independence from each others" (McIntosh 1969)

1 Introduction

Recent interest among prominent economic theorists for the issue of self-control (see e.g., Gul and Pesendorfer (2001, 2004, 2005), Fudenberg and Levine (2006, 2010)), often builds on the hypothesis that an individual may be better described by a multiplicity of selves who may have diverging interests and intentions than as a single piece of coherent intentions. Various ways to model those selves and interaction between them have recently been investigated. Often they amount to enriching the standard model by adding short-run impatient selves. In this paper, we argue that the quantum approach to decision-making provides a suitable framework to the McIntosh's paradox of self-control because the indeterminacy of individual preferences precisely means multiplicity of the selves (the potential eigentypes).

The quantum approach to decision-making and to modelling behavior more generally ((see e.g., Deutsch (1999), Busemeyer et al. (2006, 2007, 2008), Danilov

D. Song et al. (Eds.): QI 2011, LNCS 7052, pp. 71–82, 2011.

et al. (2008), Franco (2007), Danilov et al. (2008), Khrennikov (2010), Lambert-Mogiliansky et al. (2009)) opens up for the issue of self-control or, as we prefer to call it self-management, as soon as we consider dynamic individual optimization. In contrast with the recent papers on self-control, we can address these issues without introducing the time dimension but focusing instead on the sequential character of decision-making. In this paper we propose an introduction to dynamic optimization using the Type indeterminacy model (Lambert-Mogiliansky et al. 2009). The basic assumption will be that the agent is aware of his type indeterminacy, that is of the way his decisions have impact on his future type and consequently on future choices and (expected) outcomes. We show that, in a TI-model, dynamic optimization translates into a game of self-management among multiple selves. Its natural solution concept is Bayes-Nash equilibrium i.e., a decentralized equilibrium among the selves.

We are used to situations where current decisions affect future decisions. This is the case whenever the decisions are substitutes or complements. A choice made earlier changes the value of future choices by making them more valuable when the choices are complements or less valuable when they are substitutes. The preferences are fixed over time but the endowment changes. The theories of addiction address the case when a current decision impact on future *preferences*.[1] Generally however, the decision theoretical literature assumes that preferences are fixed unless a special additional structure is provided. When it comes to dynamic optimization, backward induction is the standard approach and it secures that final decisions are consistent with initial plans. There is now considerable evidence from experimental economics and psychology that people are dynamically inconsistent. There exists also a vast theoretical literature pioneered by Strotz (1955) dealing with various type of time inconsistency (see also Machina, 1989, Sarrin and Wakker,1998). A large share of this literature has focused on inconsistency that arises because the individual does not discount the future at a constant rate. Some form of myopia is assumed instead (e.g., quasi-hyperbolic discounting). Dynamic inconsistency has also been exhibited in experiments with sequences of choices but no discounting (Busemeyer et al., 2000, Hey and Knoll, 2007, Cubitt, Starmer and Sugden, 1998). For example Busemeyer and Barkan (2003) presented decision makers with a computer controlled two stage gamble. Before playing and knowing the outcome of the first stage, the person made plans for the choice on the second stage depending on each possible outcome of the first stage. Subsequently, the first stage was played out and the person was then given an opportunity to change her choice for the second stage game after observing the first stage outcome. The results demonstrated a systematic form of dynamic inconsistency that cannot be explained appealing to time preferences. In this paper we are dealing with (apparent) dynamic inconsistency that arises in the absence of any discounting.

---

[1] Consuming drugs today makes you more willing to consume tomorrow and you may end up as a drug addict. Knowing that, a rational agent may refrain from an even small and pleasant consumption today in order not to be trapped in addiction.

In a Type Indeterminacy context, preferences are indeterminate and therefore they change along with the decisions that are made. The person(type) who makes the first decision is not the same as the person who makes the second decision, it is not surprising that the two decisions are not consistent with each other.[2] They simply do not arise from the same preferences. Therefore, some instances of "apparent" dynamic inconsistency are to be expected. But does this mean that we must give up all idea of consistency and of dynamic optimization? Of course not.[3] The dynamics of the change in preferences in any specific TI-model are well-defined. An individual who is aware of how his decision today affects his preferences tomorrow will simply integrate this feature in his optimization problem. For instance Bob may very well be aware (as we assume in our lead example) of the fact that when he is in a calm mood because e.g., he took a decision that involves no risk, he also usually finds himself in a rather empathetic mood. In contrast, when taking a risky decision, he is tense and tends to behave egoistically. That awareness may prompt a decision with respect to risk-taking that is aimed at controlling his future mood(type) in order to achieve an overall higher utility.[4] In the last section we argue that what may be perceived as a dynamically inconsistent behavior need not be. Instead, it may reflect the rational reasoning of a type indeterminate agent.

Closely related to this paper is one, earlier mentioned, articles by Fudenberg and Levine (2006). They develop a dual self model of self-control that can explain a large variety of behavioral paradoxes. In their model there is a long-term benevolent patient self and a multiplicity of impulsive short-term selves - one per period. This particular structure allows them to write the game as a decision problem. In contrast, we are dealing with a full-fledged game involving a multiplicity of simultaneous (symmetric) selves in each period. All selves are equally rational and care about the future expected utility of the individual. The dual self model is designed to capture the management of impatience and it has a strong predictive power. Interestingly, both the dual self model and the TI-model can show that (apparent) dynamic inconsistency may arise as a result of rational self-control. We trust that the quantum approach has the potential to capture self-management issues reflecting a wide range of conflicting interests within the individual. The TI self-management approach is also related to another line of research belonging to Benabou and Tirole (2011). In a recent paper the propose a theory of identity management which bears interesting similarities with ours. Benabou and Tirole do not have a multiplicity of selves but as in

---

[2] Yukalov and Sornette (2010) have proposed that this type of dynamic inconsistency can be explained by quantum models of decision making. But they are not interested in the issue of optimization.

[3] Another approach is to use the hypothesis of type indeterminacy to develop a theory of bounded rationality. The assumption would be that individual preferences change but the agent is not aware of that. We believe that a first step is to maintain the rationality assumption and investigate the implications of type indeterminacy.

[4] In a experimental paper,"Your Morals Might Be Your Mood" the authors (Kirsteiger et al. 2006.) show how the mood (induced by a film sequence) determines preferences in a next following fully unrelated gift exchange game.

the TI-model the agent is "what he does" and so a reason for making choice is to determine who he is with respect to next period's action. In their model the reason is that the agent does not know his deep preferences, learns but keeps forgetting about it.

2   Dynamic Single Player Optimization

Let us consider a series of two decisions in an ordered sequence. Firs, the agent makes her choice of one option in $\{a_1, ..., a_n\}$ referred to as called DS1 (Decision Situation 1) and thereafter of one option in $\{x_1, ..., x_n\}$ referred to as DS2. Generally the utility value of the x-choice may depend on the choice of the a-option. This is the case when the two decisions are to some extent complementary or substitute. Here we shall assume that the two choices are independent. One example that we investigate later as an illustration is when the first decision situation concerns a portfolio of financial assets and the second how to spend the evening with your spouse. This assumption of independence is made to exhibit in the simplest possible context the distinctions between the predictions about behavior in the classical and respectively the type indeterminacy model of decision-making.

The agent is characterized by her preferences, that is an ordering of the different options. We can distinguish between $n!$ possible orderings called θ_i (or $a-$type) relevant to the $a-$choice and similarly, $n!$ different types τ_i relevant to the $x-$choice. There is no discounting so the utility of the two-period decision problem can be written as the utility of the first period (i.e., from the $a-$choice) plus the utility of the second period (i.e., from the $x-$choice):

$$U(a_i, x_i) = U(a_i; t_0) + U(x_i; t_1)$$

where t_0 is the type of the agent i.e., her preferences with respect to both choices (a and x) at time $t = 0$ and t_1 is the type of the agent after her first decision at time $t = 1$. The optimization problem generally writes:

$$\max_{\{a_1, ...a_n\} \times \{x_1, ..., x_n\}} [U(.; t_0) + U(.; t_1)].$$

The classical model
For the case the agent is classical, all type characteristics are compatible with each other and the set of possible types is $\{\theta_1, ..., \theta_{n!}\} \times \{\tau_1, ..., \tau_{n!}\}$. It has cardinality $(n!)^2$ and elements $\theta_i \tau_j$, $i = 1, ..n$, $j = 1, ...n$. Moreover $t_0 = t_1$ since nothing happens between the two choices that could affect the preferences of the agent. The agent knows her type which is a priory determined. The optimization problem is fully separable and writes

$$\max_{\{a_1, ...a_n\} \times \{x_1, ..., x_n\}} U(a_i, x_i) = \max_{\{a_1, ..., a_n\}} U(a_i; t_0) + \max_{\{x_1, ..., x_n\}} U(x_i; t_0)$$

This is the simplest case of dynamic optimization, it boils down to two static optimization problems.

The *Type Indeterminacy model*

In the TI-model, a decision-maker is represented by his state or type (the two terms will be used interchangeably) which captures his preferences. A type is a vector $|t_i\rangle$ in a Hilbert space. A simple decision situation (DS) is represented by an (linear) operator.[5] The act of choosing in a decision situation actualizes an eigentype[6] of the operator (or a superposition[7] of eigentypes if more than one eigentype would make the observed choice). An eigentype is information about the preferences (type) of the agent. For instance consider a model where the agent has preferences over sets of three items, i.e. he can rank any 3 items from the most preferred to the least preferred. Any choice experiment involving three items is associated with six eigentypes corresponding to the six possible ranking of the items. If the agent chooses a out of $\{a, b, c\}$ his type is projected onto some superposition of the ranking $[a > b > c]$ and $[a > c > b]$. The act of choosing is modelled as a measurement of the (preference) type of the agent and it impacts on the type i.e., it changes it (for a detailed exposition of the TI-model see Lambert-Mogiliansky et al. 2009).

We know (see Danilov et Lambert-Mogiliansky 2008) that there is no distinction with the classical (measurement) analysis when the two DS commute. Therefore we shall assume that DS1 and DS2 are *non-commuting* operators which means that the type characteristics θ and τ are incompatible or equivalently that the relevant set of type is $\{\theta_1, ..., \theta_{n!}\} \cup \{\tau_1, ..., \tau_{n!}\}$ with cardinality $2n!$. When dealing with non-commuting operators we know that the order of decision-making matters. The operator DS1 acts on the type of the agent so the resulting type t_1 is a function of a. Without getting into the details of the TI-model (which we do in the next section) we note for that optimization problems writes

$$\max_{\{a_1, ... a_n\} \times \{x_1, ..., x_n\}} [U(a_i; t_0) + U(x_i; t(a))]$$

So we see that the two decision situations are no longer separable. When making her first decision the rational agent takes into account the impact on his utility in the second decision situation as well.

We shall below investigate an example that illustrates the distinction between the two optimization problems and suggest that the type indeterminacy model captures realistic features of human behavior that can only be captured with additional structure in a classical model.

2.1   An Illustrative Example

We have one agent and we call him Bob. Bob who just inherited some money from his aunt, faces two consecutive decisions situations DS1:$\{a_1, a_2\}$ and DS2: $\{x_1, x_2\}$. For the sake of concreteness, the first decision is between buying state

---

[5] In Physics such measurement operators are called "observables".

[6] The eigentypes are the types associate with the eigenvalues of the operator i.e., the possible outcomes of the measurement of the *DS*.

[7] A superposition is a linear combination of the form $\sum \lambda_i |t_i\rangle$; $\sum \lambda_i^2 = 1$.

obligations (a_1) or risky assets (a_2). The second choice decision is between a stay at home evening (x_1) or taking his wife to a party (x_2). The relevant type characteristics to DS1 have two values (eigentypes): cautious (θ_1) risk loving (θ_2). In DS2 the type characteristics has two values as well: (τ_1) egoistic versus generous/empathetic (τ_2).

We belowdefine the utility associated to the different choices. The most important to keep in mind is that in DS2 the generous/empathetic type experiences a high utility when he pleases his wife. The egoist type experience a low utility from the evening whatever he does but always prefers to stay home.

Classical optimization. Let us first characterize the set of types. Since both type characteristics each have two values, Bob may be any of the following four types $\{\theta_1\tau_1, \theta_1\tau_2, \theta_2\tau_1, \theta_2\tau_2\}$.

The utility is described by table 1 and 2 below

Tab. 1
a_1	a_2
$U(a_1; \theta_1\tau_1) = U(a_1; \theta_1\tau_2) = 4$	$U(a_2; \theta_1\tau_1) = U(a_2; \theta_1\tau_2) = 2$
$U(a_1; \theta_2\tau_1) = U(a_1; \theta_2\tau_2) = 2$	$U(a_2; \theta_2\tau_1) = U(a_2; \theta_2\tau_2) = 3$

so only the θ value matters for the a−choice.

Tab.2
x_1	x_2
$U(x_1; \theta_1\tau_1) = U(x_1; \theta_2\tau_1) = 2$	$U(x_2; \theta_1\tau_1) = U(x_2; \theta_2\tau_1) = 0$
$U(x_1; \theta_1\tau_2) = U(x_1; \theta_2\tau_2) = 1$	$U(x_2; \theta_1\tau_2) = U(x_2; \theta_2\tau_2) = 8$

so here only the τ value matters for the x−choice.

The tables above give us immediately the optimal choices:

$\theta_1\tau_1 \rightarrow (a_1, x_1)$	$\theta_2\tau_1 \rightarrow (a_2, x_1)$
$\theta_1\tau_2 \rightarrow (a_1, x_2)$	$\theta_2\tau_2 \rightarrow (a_2, x_2)$

Using the values in table 1 and 2, we note that type $\theta_1\tau_2$ achieves the highest total utility of 12. the lowest utility is achieved by $\theta_2\tau_1$.[8] While Bob knows his type, we do not. We know that "the population of Bobs" is characterized by the following distribution of types:

$\theta_1\tau_1 \rightarrow 0.15$	$\theta_2\tau_1 \rightarrow 0.35$
$\theta_1\tau_2 \rightarrow 0.35$	$\theta_2\tau_2 \rightarrow 0.15$

We note that the distribution of types in the population of Bobs exhibit a statistical correlation between the θ and τ type characteristics.

---

[8] Note that we here assume that we can compare the utility of the different types of Bob. This goes beyond standard assumption in economics that preclude inter personal utility comparisons. But is in line with inter personal comparisons made in the context of social choice theory.

2.2 A TI-model of Dynamic Optimization

By definition the type characteristics relevant to the first DS1 is $\theta, \theta \in: \{\theta_1, \theta_2\}$. Subjecting Bob to the a−choice is a measurement of his θ characteristics. The outcome of the measurement maybe θ_1 or θ_2 and Bobs collapses on an eigentype or the outcome may be null (when both θ_1 and θ_2 choose the same action).[9] The type characteristics relevant to DS2 is $\tau, \tau \in \{\tau_1, \tau_2\}$. Since the two DS do not commute we can write

$$|\theta_1\rangle = \alpha_1 |\tau_1\rangle + \alpha_2 |\tau_2\rangle$$
$$|\theta_2\rangle = \beta_1 |\tau_1\rangle + \beta_2 |\tau_2\rangle$$

where $\alpha_1^2 + \alpha_2^2 = 1 = \beta_1^2 + \beta_2^2$. For the sake of comparison between the two models we let $\alpha_1 = \beta_2 = \sqrt{.3}$ and $\alpha_2 = \beta_1 = \sqrt{.7}$. Bob's initial type or state is

$$|t\rangle = \lambda_1 |\theta_1\rangle + \lambda_2 |\theta_2\rangle, \ \lambda_1^2 + \lambda_2^2 = 1$$

with $\lambda_1 = \lambda_2 = \sqrt{.5}$.

When discussing utility in a TI-model one should always be careful. This is because in contrast with the classical model, there is not one single "true type" who evaluates the utility value of all choice options. A key assumption is (as in TI-game see Lambert-Mogiliansky 2010) that *all the reasoning of the agent is made at the level of the eigentype* who knows his preferences (type), has full knowledge of the structure of the decision problem and cares about the expected payoff of Bob's future incarnations (type). The utility value for the current decision is evaluated by the eigentype who is reasoning. So for instance when Bob is of type t, two reasonings take place. One performed by the θ_1 eigentype and one performed by θ_1 eigentype. The θ−types evaluate the second decision, using the utility of the type resulting from the first decision. The utility of a superposed type is the weighted average of the utility of the eigentypes where the weights are taken to be the square of the coefficient of superposition.[10] The utility of the eigentypes are depicted in the table 3 and 4 below

Tab. 3
$U(a_1; \theta_1) = 4$	$U(a_2; \theta_1) = 2$
$U(a_1; \theta_2) = 2$	$U(a_2; \theta_2) = 3$

, and Tab. 4
$U(x_1; \tau_1) = 2$	$U(x_2; \tau_1) = 0$
$U(x_1; \tau_2) = 1$	$U(x_2; \tau_2) = 8$

.

As earlier noted Bob in state t performs two (parallel) reasonings. We proceed by backward induction to note that trivially since the "world ends after DS2", τ_1 chooses x_1 and τ_2 chooses x_2 (as in the classical model). We also note that:

---

[9] More correctly when both our eigentypes choose the same action in DS1, DS1 is a null measuremnt i.e., it does not allows to distinguish between the eigentypes.

[10] We note that in the TI-model we cannot escape inter type utility comparison. We must aggregate the utilies over different selves to compute the optimal decisions. However just as in social choice theory there is no unique way of aggregating individual utility into a social value. We return this issue in the discussion.

$U(x_1; \tau_1) = 1 < U(x_2; \tau_2) = 8$. The τ_2 incarnation of Bob always experiences higher utility than τ_1.

The TI-model has the structure of a two-stage maximal information[11] TI-game as follows. The set of players is $N : \{\theta_1, \theta_2, \tau_1, \tau_2\}$, the θ_i have action set $\{a_1, a_2\}$ they play at stage 1. At stage 2, it is the τ_i players' turn, they have action set $\{x_1, x_2\}$. There is an initial state $|t\rangle = \lambda_1 |\theta_1\rangle + \lambda_2 |\theta_2\rangle$, $\lambda_1^2 + \lambda_2^2 = 1$ and correlation between players at different stages: $|\theta_1\rangle = \alpha_1 |\tau_1\rangle + \alpha_2 |\tau_2\rangle$ and $|\theta_2\rangle = \beta_1 |\tau_1\rangle + \beta_2 |\tau_2\rangle$. The utility of the players is as described in tables 3 and 4 when accounting for the players' concern about future selves. So for a θ−player, the utility is calculated as the utility from the choice in DS1 plus the expected utility from the choice in DS2 where expectations are determined by the choice in DS1 as we shall see below.

The question is how will Bob choose in DS1, or how do his different θ-eigentype or selves choose? We here need to do some simple equilibrium reasoning.[12] Fix the strategy of pure type θ_1, say he chooses "a_1".[13] What is optimal for θ_2 to choose? If he chooses "a_2" the resulting type after DS1 is $|\theta_2\rangle$. The utility, in the first period, associated with the choice of "a_2" is $u(a; \theta_2) = 3$. In the second period Bob's type is $|\theta_2\rangle = \beta_1 |\tau_1\rangle + \beta_2 |\tau_2\rangle$ which, given what we know about the optimal choice of τ_1 and τ_2, yields an expected utility of $\beta_1^2 [U(x_1; \tau_1) = 1] + \beta_2^2 [U(x_2; \tau_2) = 8] = .7 + 8(.3) = 3.1$. The total (for both periods) expected utility from playing "a_2" for θ_2 is

$$EU(a_2; \theta_2) = 3 + 3.1 = 6.1$$

This should be compared with the utility, for θ_2, of playing "a_1" in which case he pools with θ_1 so the resulting type in the first period is the same as the initial type i.e., $|t\rangle = \lambda_1 |\theta_1\rangle + \lambda_2 |\theta_2\rangle$. The expected utility of playing a_1 is $u(a_1; \theta_2) = 2$ in the first period plus the expected utility of the second period. To calculate the latter, we first express the type vector $|t\rangle$ in terms of $|\tau_i\rangle$ eigenvectors:

$$|t\rangle = \lambda_1 (\alpha_1 |\tau_1\rangle + \alpha_2 |\tau_2\rangle) + \lambda_2 (\beta_1 |\tau_1\rangle + \beta_2 |\tau_2\rangle)$$
$$= (\lambda_1\alpha_1 + \lambda_1\beta_1) |\tau_1\rangle + (\lambda_1\alpha_2 + \lambda_2\beta_2) |\tau_2\rangle.$$

---

[11] Maximal information TI-game are the non-classical counter-part of classical complete information games. But in a context of indeterminacy, it is not equivalent to complete information because there is an irreducible uncertainty. It is impossible to know all the type characteristics with certainty.

[12] Under equilibrium reasoning, an eigentype is viewed as a full valued player. He makes assumption about other eigentypes' play at difference stages and calculate his best reply to the assumed play. Note that no decision is actually made so no collapse actually takes place. When he finds out what is optimal for him, he checks whether the assumed play of others is actually optimal for them given his best response. We have an equilibrium when all the eigentypes are best responding to each others.

[13] We note that the assumption of "a_1" is not fully arbitrary since a_1 gives a higher utility to θ_1 than a_2. However, we could just as well have investigated the best reply of θ_1 after fixing (making assumption) the choice of θ_2 to a_2. See further below and note 12 for a justification of our choice.

The second period's expected utility is calculated taking the optimal choice of τ_1 and τ_2:

$$\left(\lambda_1^2\alpha_1^2 + \lambda_2^2\beta_1^2 + 2\lambda_1\alpha_1\lambda_2\beta_1\right)1 + \left(\lambda_1^2\alpha_2^2 + \lambda_2^2\beta_2^2 + 2\lambda_1\alpha_2\lambda_2\beta_2\right)8 = 0.959 + 7.669$$
$$= 8.63.$$

which yields

$$EU\left(a_1; \theta_2\right) = 2 + 8,63 = 10,63 > EU\left(a_2; \theta_2\right) = 3 + 3.1 = 6.1$$

So we see that there is a gain for θ_2 of preserving the superposition i.e., it is optimal for pure type θ_2 to forego a unit of utility in DS1 and play a_1 (instead of a_2 as in the classical model). It can also be verified that given the play of θ_2 it is indeed optimal for θ_1 to choose a_1. The solution to dynamic optimization is an "inner" Bayes-Nash equilibrium where both θ_1 and θ_2 to play a_1.[14]

The interpretation is that Bob's θ_2 type understands that buying risky assets appeals to his risk-loving self which makes him tense. He knows that when he is tense, his egoistic self tends to take over. So, in particular, in the evening he is very unlikely to feel the desire of pleasing his wife - his thoughts are simply somewhere else. But Bob also knows that when he is in the empathetic mood i.e., when he enjoys pleasing his wife and he does it, he always experiences deep happiness. So his risk-loving self may be willing to forego the thrill of doing a risky business in order to increase the chance for achieving a higher overall utility.

Multiple-selves, individual management and dynamic inconsistency. This paper is offering a new perspective on self-management that emerges from type indeterminacy in a dynamic optimization context. By construction the outcome exhibits no inconsistency. On the contrary Bob is a self-aware rational agent. Yet, we shall argue that our approach may provide some new insights with respect to the issue of dynamic inconsistency.

The model has been designed to exhibit distinctions between classical and TI optimization in the simplest possible context i.e., when the two decisions are independent and in the absence of discounting. This corresponds to the gambling example discussed in the introduction. The decisions in the two gambles can be viewed as independent. Moreover the inconsistency is between the declared intentions (plans) and the actual choices is not due to time discounting since we have none. If we do, as in the described experiment, ask Bob about his plans i.e., what he prefers to do before actually making any decision, our example will exhibit a similar instance of "dynamic inconsistency". Assume that we have a population of "Bobs", initially in a (superposed) state. When asked what he likes to do with the portfolio, Bob will answer with some probability that he wants to enjoy the thrill of risky business. When asked further what

---

[14] The equilibrium need not be unique. A similar reasoning could be made for both θ−type pooling on a_2. The inner game is a coordination game. It make sense to assume that coordination is indeed achieved since all the reasoning occurs in one single person.

he plans to do in the evening, he will with some probability answer that he wants to please his wife.[15] Note first that these responses are sincere because a significant "part" of Bob enjoys risk and he knows that he can be very happy when his wife also is happy. However when the time comes for actually making the portfolio decision, we observe that the agents always choose non risky assets (they buy state obligations). This is inconsistent with the declared intentions. Indeed it seems in contradiction with the preferences sincerely revealed to the experimentalist. However, we argue that this apparent inconsistency may hide a quite sophisticated self-management calculation. The agent is aware that he is constrained by the dynamics of type indeterminacy. He would like to enjoy the excitement of risk and the pleasure of shared happiness but he knows that it is very unlikely that he will be able to appreciate both. Therefore, he chooses to increase the chance for securing his ability to enjoy his wife's happiness at the cost of the excitement of risk. So in fact he is not being inconsistent at all, not even with his initially revealed preferences. Here apparent inconsistency is due to the fact that the outside observer makes the incorrect assumption that Bob has fixed preferences. In that case there would be no issue of self-management but simply of maximizing utility and the observed behavior would indeed be dynamically inconsistent. So we propose that some instance of (apparent) dynamic inconsistency maybe explained by a rational concern for self-management.

3   Concluding Remarks

In this paper, we proposed an introduction to dynamic optimization for Type Indeterminate agents. Our model is that of a rational agent aware of his own indeterminacy. We found that type indeterminacy has very interesting implications in terms of self-management. Dynamic decision-making becomes a non trivial game between the multiple potential eigentypes(selves) of the individual. The outcome is a Bayes-Nash equilibrium among the potential selves. In the example that we investigate it delivers predictions that make a lot of sense in terms of self-control and self-management. When complemented with a preliminary question about preferences, the equilibrium features apparent dynamic inconsistency in the absence of any time discounting. One distinctive feature of our approach is that while many models of self control do rely on the multiplicity of selves, they often assume some asymmetry so one of the selves dominates e.g., the long-term self in Fudenberg and Levine (2006) or the current self in other models. The decentralized equilibrium approach that emerges from the TI-model does not feature any asymmetry between the selves such that it singles out one particular self as the dominant one. Yet, we obtain self control. This is because indeterminacy in itself generates the issue of self-management.[16]

---

[15] We do not discuss the question as to whether simply responding to a question has an impact on Bob's type i.e., forces a collapse. The argument is equally valid but requires some further specification when questionning affect the state.

[16] Although we have not done it, the TI-model does allow to account for asymmetries for instance the eigentypes associated with the first period DS may be the only forward-looking selves.

References

1. Barkan, R., Busemeyer, J.R.: Modeling dynamic inconsistency with a changing reference point. Journal of Behavioral Decision Making 16, 235–255 (2003)
2. Benabou, R., Tirole, J.: Identity, Morals and Taboos: Beliefs as Assets. Quaterly Journal of Economics (2011) (forthcoming)
3. Busemeyer, J.R., Weg, E., Barkan, R., Li, X., Ma, Z.: Dynamic and consequential consistency of choices between paths of decision trees. Journal of Experimental Psychology: General 129, 530–545 (2000)
4. Busemeyer, J.R., Wang, Z., Townsend, J.T.: Quantum Dynamics of Human Decision-Making. Journal of Mathematical Psychology 50, 220–241 (2006)
5. Busemeyer, J.R.: Quantum Information Processing Explanation for Interaction between Inferences and Decisions. In: Proceedings of the Quantum Interaction Symposium AAAI Press, Menlo Park (2007)
6. Busemeyer, J.R., Santuy, E., Lambert-Mogiliansky, A.: Distinguishing quantum and markov models of human decision making. In: Proceedings of the the Second Interaction Symposium, QI 2008, pp. 68–75 (2008a)
7. Busemeyer, J.R., Lambert-Mogiliansky, A.: An Exploration of Type Indeterminacy in Strategic Decision-Making. In: Bruza, P., Sofge, D., Lawless, W., van Rijsbergen, K., Klusch, M. (eds.) QI 2009. LNCS(LNAI), vol. 5494, pp. 113–128. Springer, Heidelberg (2009)
8. Cubitt, R.P., Starmer, C., Sugden, R.: Dynamic choice and the common ratio effect: An experimental invesigation. Economic Journal 108, 1362–1380 (1998)
9. Danilov, V.I., Lambert-Mogiliansky, A.: Measurable Systems and Behavioral Sciences. Mathematical Social Sciences 55, 315–340 (2008)
10. Danilov, V.I., Lambert-Mogiliansky, A.: Decision-making under non-classical uncertainty. In: Proceedings of the the Second Interaction Symposium (QI 2008), pp. 83–87 (2008)
11. Danilov, V.I., Lambert-Mogiliansky, A.: Expected Utility under Non-classical Uncertainty. Theory and Decision 2010/68, 25–47 (2009)
12. Deutsch, D.: Quantum Theory of Propability and Decisions. Proc. R. Soc. Lond. A 455, 3129–3137 (1999)
13. Franco, R.: The conjunction Fallacy and Interference Effects (2007), arXiv:0708.3948v1
14. Franco, R.: The inverse fallacy and quantum formalism. In: Proceedings of the Second Quantum Interaction International Symposium (QI 2008), pp. 94–98 (2008)
15. Fudenberg, Levine: A Dual Self Model of Impulse Control. American Economic Review 96, 1449–1476 (2006)
16. Fudenberg, Levine.: Timing and Self-Control working paper (2010)
17. Gul, F., Pesendorfer, W.: Temptation and Self Control. Econometrica 69, 1403–1436 (2001)
18. Gul, F., Pesendorfer, W.: Self Control and the Theory of Consumption. Econometrica 72, 110–158 (2004)
19. Gul, F., Pesendorfer, W.: Self Control, Revealed Preference and Consumption Choice. Review of Economic Studies (2005)
20. Hey, J.D., Knoll, J.A.: How far ahead do people plan? Economic Letters 96, 8–13 (2007)
21. Khrennikov, A.: Ubiquitous Quantum Structure - From Psychology to Finance. Springer, Heidelberg (2010)

22. Kirsteiger, G., Rigotti, L., Rustichini, A.: Your Morals Might be Your Moods. Journal of Economic Behavior and Organization 59/2, 155–172 (2006)
23. Lambert-Mogiliansky, A., Zamir, S., Zwirn, H.: Type indeterminacy - A Model of the KT(Khaneman Tversky)- man. Journal of Mathematical Psychology 53/5, 349–361 (2009)
24. Lambert-Mogiliansky, A.: Endogenous preferences in games with Type-Indeterminate Players, FS 10-08, pp. 70–77. AAAI Press, Menlo Park (2010)
25. La Mura, P.: Correlated Equilibria of Classical strategies with Quantum Signals. International Journal Of Quantum Information 3, 183–188 (2005)
26. La Mura, P.: Prospective Expected Utility. In: Proceedings of the the Second Quantum Interaction International Symposium (QI 2008), pp. 87–94 (2008)
27. Machina, M.: Dynamic inconsistency and non-expected utility models of choice under uncertainty. Journal of Economic Literature 27, 1622–1668 (1989)
28. Strotz, R.H.: Myopya and Time Inconsistency in Dynamic Utility Maximization. Review of Economic Studies 23(3), 165–180 (1956)

Pseudo-classical Nonseparability and Mass Politics in Two-Party Systems

Christopher Zorn[1] and Charles E. Smith[2]

[1] Department of Political Science, Pennsylvania State University, University Park, PA
[2] Department of Political Science, University of Mississippi, Oxford, MS

Abstract. We expand the substantive terrain of QI's reach by illuminating a body of political theory that to date has been elaborated in strictly classical language and formalisms but has complex features that seem to merit generalizations of the problem outside the confines of classicality. The line of research, initiated by Fiorina in the 1980s, seeks to understand the origins and nature of party governance in two-party political systems wherein voters cast partisan ballots in two contests, one that determines partisan control of the executive branch and another that determines party control of a legislature. We describe how research in this area evolved in the last two decades in directions that bring it now to the point where further elaboration and study seem natural in the more general formalistic and philosophical environments embraced in QI research. In the process, we find evidence that a restriction of a classical model that has animated work in the field appears violated in a form that leads one naturally to embrace the superposition principle. We then connect classical distinctions between separable and nonseparable preferences that are common in political science to their quantum and quantum-like counterparts in the QI literature, finding special affinity for a recently-introduced understanding of the distinction that provides a passageway into the boundary between fully quantum and fully classical views of the distinction and thereby provides new leverage on existing work germane to the theory.

1 Introduction

Among all of the academic specialties customarily identified as social sciences, political science is perhaps the greatest "debtor" discipline, in the sense that so many of the theories and methods and models put to the task of understanding politics are borrowed from scholars working in other fields. It is thus predictable that some of the latest and most promising theoretical and methodological innovations providing insight into the operation of politics are not native to political science. What is surprising is their footing in quantum mechanics. Long thought in the main to be a theory with applications exclusive to the realm of the near-unobservably small, where probabilities rather than observable mechanics propagate in accordance with causal laws, the 21st century is becoming witness to an ever-growing export market for the quantum formalisms and the probability theory native to them. This paper follows that trend by illuminating a body of political theory that to date has been elaborated in strictly classical language and formalisms but has complex features that seem to merit generalizations of the problem outside the confines of strict classicality.

D. Song et al. (Eds.): QI 2011, LNCS 7052, pp. 83–94, 2011.

2   Balancing Theory

In the U.S., one of the most prominent strands of research on the origins of party governance was initiated by Fiorina in the late 1980s and early 1990s [1]. In contrast to classical, Downsian [2] models, where voters with policy preferences that are more moderate than the positions staked out by parties in two-party systems choose (if possible) the closest of the two alternatives, Fiorina's thesis emphasizes the importance of the two institutional choices in U.S. politics: the Congress and the presidency. In his model, voter desires for moderation can be realized by "splitting the ticket" – voting for the Republican candidate in one institutional choice setting and the Democrat in the other. Likewise, voters with more extreme positions can maximize their returns by choosing one party across both institutional contests. This strand of research thus contrasts with the binary choice (Democrat (D) versus Republican (R)) tradition from Downs by framing the problem as a choice set for party governance (G) across four mutually exclusive options, $G = [D_E D_L, D_E R_L, R_E D_L, R_E R_L]$, where the subscripts distinguish the election contesting control of the executive branch from the one deciding control of the legislature.

Fiorina's initial formulation of the problem defined the choice options and voter positions relative to them in a one-dimensional, policy-specific, liberal-versus-conservative Euclidean space.[1] Across individuals, different issues have different levels of salience; moreover, individual understandings/predictions of where the parties stand on issues may be variable. For one or all of these reasons, measured policy preferences in the mass public are not stable across time, an empirical regularity traceable back at least to Converse [3]. Another complication is that, a priori, the universe of salient policies in an election is difficult to determine, and thus measure, for all voters/respondents/subjects. Given all these givens, it is perhaps not surprising that many of the scholars who have investigated the empirical relevance of Fiorina's "policy balancing" theory report that it provides little or no observable, explanatory purchase to our understanding of partisan or bipartisan (i.e. ticket-splitting) choice [4,5].

However, "party balancing" is a different matter. As explained in [6], "the act of 'policy balancing' implies that individual voters ultimately engage in 'party balancing,'" a process whereby voters adjust their preferences regarding which party should control one institution based on either preferences for or expected outcomes about partisan control of the other. The focus of this study was narrow: the authors took as their primary task an analysis of how then-customary, statistical models of candidate/party choice in U.S. Congressional elections might be better specified given an account of measurement metric implications derivative of one (of several) possible, theoretical exposition(s) of party balancing. However, both the theory underlying the hypotheses tested in the research and the data used to do so are perhaps of broader interest. On the theoretical side, this study leans on one account of how social scientists understand the distinction between preference separability and nonseparability, issues that merit attention given their kinship (and lack thereof) with the quantum mechanical meanings of those terms. On

---

[1] A generalization of that model to N dimensions is straightforward, but specifications of its empirical implications relative to the four partisan choice options are not easily defined in a parsimonious fashion.

the empirical side, results in the survey data used in [6] are not readily accommodated by classical formalizations. In the subsections that follow, we elaborate.

2.1 Classical Views of Separability and Nonseparability

Social scientists understand and use the words *separable* and *nonseparable* in ways that are distinct from the quantum mechanical meanings of the terms. The most general tradition uses the terms to distinguish between two types of preference orders. The most basic, classical example is one voter with two considerations, observable as two bits. Preferences are said to be separable when each of those preferences arises independent of the consideration of or outcome on the other. Of the twenty-four (4!) possible preference orderings in the two-bit example, the eight orderings with last preferences as mirror images of the first (e.g. 00 01 10 11) are understood to be separable orderings when the considerations are of equal salience and the orderings are observed across groups of voters as invariant to the order in which the preferences are measured [7].

A visually intuitive alternative to understanding separability and nonseparability in previous social science work (including that on balancing theory) is animated by a simple model and illustration. For Figure 1, define S as an initial state belief vector that can be used to describe considerations over preferences regarding partisan control (Republican versus Democratic) of both the executive and the legislative branches in an election. Belief vectors regarding partisan options in the two-dimensional space can then be described in terms of coordinates specific to each branch. Further, define a simple Euclidean distance in the space:

$$||S_E - S_L||_{\mathbf{I}} = \sqrt{(R_E - D_E)^2 + (R_L - D_L)^2} \tag{1}$$

with

$$\mathbf{I} = \begin{bmatrix} \omega_{11} & \omega_{12} \\ \omega_{21} & \omega_{22} \end{bmatrix} \tag{2}$$

which can be interpreted as weights. Specifically, the main diagonal weights signify the salience of the two, associated dimensions of party governance, their ratio the relative importance of them. In order for the space to remain Euclidean, the off diagonal elements must be equal; when they are jointly equal to zero, "there is no interaction between" [8] the considerations, and the preferences arising from them are said classically to be separable.

Now consider an alternative transition matrix:

$$\mathbf{A} = \begin{bmatrix} \alpha_{11} & \alpha_{12} \\ \alpha_{21} & \alpha_{22} \end{bmatrix} \tag{3}$$

with the restriction $\alpha_{12} = \alpha_{21} = \alpha$. Replacing $\mathbf{I}$ with $\mathbf{A}$ gives:

$$||S_E - S_L||_{\mathbf{A}} = \sqrt{\alpha_{11}(R_E - D_E)^2 + 2\alpha(R_E - D_E)(R_L - D_L) + \alpha_{22}(R_L - D_L)^2} \tag{4}$$

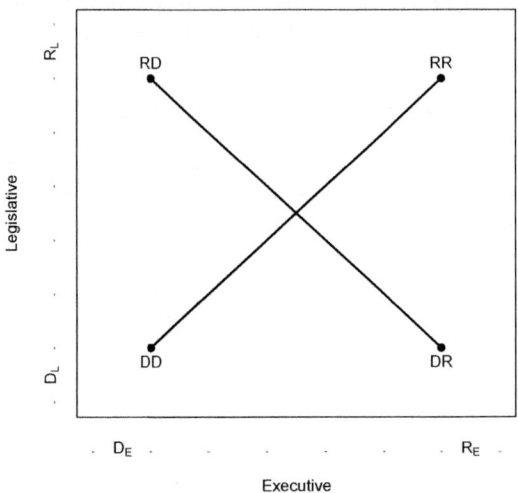

Fig. 1. Divided vs. Unified Control of Government

This is a stylized version of the weighted Euclidean norm developed by Enelow and Hinich [8].[2] Fixing coordinates at the poles of one dimension and differentiating the square to invoke preferences in the other when $\mathbf{A}$ is a matrix of positive ones gives $R_E D_L$ and $D_E R_L$. These are the choice options of the balancer – the voter who prefers a form of coalition government to one-party control. Conversely, preferences in the unifying regime – $R_E R_L$ and $D_E D_L$ – are given by fixing the off-diagonal elements of $\mathbf{A}$ at -1. Classically, these conditions imply that preferences are fully nonseparable and assume that the relevance to party governance of the executive and legislature are equal. The probabilities of the four outcomes for partisan control of government thus sum to unity when given $\pm\alpha$ and a partisan outcome in one dimension. That is, for a 4×1 state vector (ψ),

$$\psi = \begin{bmatrix} \psi DD \\ \psi DR \\ \psi RD \\ \psi RR \end{bmatrix} \longrightarrow \psi_{-\alpha} = \frac{1}{2}\begin{bmatrix} 1 \\ 0 \\ 0 \\ 1 \end{bmatrix}, \tag{5}$$

the vector transitions for voters with $-\alpha$ to the unifying regime and $\Pr(DD \mid D) = \Pr(RR \mid R) = .5$. Likewise, ψ transitions to the balancing regime and $\Pr(RD \mid D) = \Pr(DR \mid R) = .5$ with positive α, as in:

---

[2] Hinich and Munger [9] generalize the norm to N dimensions. Spatial voting theory more generally is built upon the early work of Davis and Hinich [10]; Gorman [11] is a fundamental work on the concept of separability, an idea he credits without specific citation to Leontief. Schwartz [12] was first to consider the problem in multiple elections. Lacy [7, 13, 14] offers more recent examples of applied and theoretical work on the separability-nonseparability distinction in political decision theory.

$$\psi = \begin{bmatrix} \psi DD \\ \psi DR \\ \psi RD \\ \psi RR \end{bmatrix} \longrightarrow \quad \psi_\alpha = \frac{1}{2} \begin{bmatrix} 0 \\ 1 \\ 1 \\ 0 \end{bmatrix}. \tag{6}$$

Most modeling and analysis in the social sciences proceed implicitly from the assumption that the observables involved in the models arise from separable considerations even when there are sound theoretical reasons to expect otherwise: economists aggregate goods in bundles that might not be separable in the minds of consumers, while political scientists do the same thing with issue preferences and voters. At best, these practices foreclose nuanced observation of potentially interesting phenomena; in the worst scenarios, they may lead to faulty inferences. As a result, it is easy to embrace theoretical and empirical work in the social sciences that make and/or test explicit assumptions about the distinction between separability and nonseparability. We are however given pause with respect to certain aspects of modeling and nomenclature conventions used routinely in social science in the course of defining and understanding the distinction.

2.2 Toward a More General Framework

Spatial representations of voting have been criticized for being overly restrictive, not least with respect to defining separable and nonseparable preferences. Lacy [13], for example, notes that in N dimensions, the model set out by [9] does not account for the possibility that sets of preferences might be nonseparable from one or more other sets. He also implies that the symmetry of the transition matrix **A** requires, given nonseparability, that each preference depend equally on the outcome relative to the other consideration. We have come to see these sorts of criticisms as wide of the mark. The first critique can be addressed simply by further generalizing the Hinich-Munger model. The second is only true if, as in our example above, it is assumed to be so; the weighted Euclidean norm certainly does not require, to reference our example, that the legislative preference be conditioned by the executive outcome when the reverse is true (formally: one of the main diagonal weights in **A** can be zero while the other is nonzero). Indeed, where others see differences in generality across the spatial and preference-order representations of separability and nonseparability in social science work, we, excepting presentation differences, see commonality. One tradition, the spatial theory, lays bare via a toy model the mechanics of the distinction; the other eschews a continuous, interval level metric; in two dimensions such as the problem above, they both distinguish in question order experiments or conditioning questions the same preference orders as exhibits of separability, the same orders as exhibiting nonseparability. Thus, at least relative to the balancing problem in two dimensions, the social science traditions are equivalent.

What the traditions in social science have most in common is their firm and exclusive footings in classicality. This is most readily illustrated in the context of the balancing problem with a simple hypothetical. Assume survey respondents are asked two conditioning questions about their preferences regarding partisan control of each of the two

Table 1. Conditional and Unconditional Preferences for Congressional Control, 1996

Unconditional Preference	Democratic Executive				GOP Executive			
	Dem.	Either	GOP	Total	Dem.	Either	GOP	Total
Democrat	188	16	98	302	258	10	34	302
Either/DK	77	26	151	254	151	31	74	256
GOP	33	9	339	381	99	14	266	379
Total	298	51	588	937	508	55	374	937

$$\chi_4^2 = 256 \ (P < 0.001); \ \hat{\gamma} = 0.72. \qquad \chi_4^2 = 290 \ (P < 0.001); \ \hat{\gamma} = 0.73.$$

institutions, and further that all respondents choose R across all four conditions. These observables would reveal one preference order for all subjects, with RR most preferred and DD least. As such, they would satisfy the inverse rule [13,14], which in two dimensions is a necessary and sufficient condition to establish separability via the preference order tradition of understanding these terms. From the standpoint of the spatial theory of voting, the off-diagonal of **A** would be presumed revealed as zero, and likewise, the preferences called separable. More fundamental, though, is that classically-trained social scientists would assume that something else is revealed in these observations, namely the probabilities that the respondents would have preferred R for each of the institutions in the absence of the conditioning questions. Axiomatically, the probability of R absent the conditions is a weighted average of the conditional probabilities gauged across the mutually exclusive and exhaustive options of the conditions. In our hypothetical, the unconditional probability of R as a revealed preference must then be unity in both dimensions.

A close reexamination of the data used in [6] gives pause against full embrace of such an axiom.[3] Following their Table 2 (p. 748), we report results in a contingency table from two conditional questions about partisan control of the U.S. Congress, one fixing a Republican victory in the presidential election and one fixing a Democratic win; both are compared to a variable measuring each respondent's "unconditional" preference over partisan control of Congress (that is, without conditioning on executive control). Results on the main diagonal of the tables thus denote respondents who answered consistently across the two conditions, and are described as characterizing voters with separable preferences. This is true by any classical standard, but we wondered about the purchase of the likewise classical assumption about preferences absent the conditions. The study includes and features two indicators proposed as such, one fashioned by the authors and called a "direct" measure, and another, "indirect" indicator that is among the most familiar measures in U.S. political science, the party identification measure developed by the authors of *The American Voter* [15] and used continuously since in the biannual U.S. National Election Studies. We look now at both unconditional preference measures

---

[3] The data for [6] came "from a pre-election telephone survey conducted by the Social Science Research Laboratory at the University of Mississippi between October 11 and November 3, 1996. The sample covered the lower forty-eight [U.S.] states and the District of Columbia. The data set contains 995 completed observations."

within the subset of respondents classified as having separable preferences based on their consistent partisan choices across the conditions and find provocative results.

Judged against the unconditional legislative preference measure developed by the authors, almost 16% of all survey respondents in the study are classified as having separable preferences using the conditional measures but offer a different preference absent the conditions. As a percentage of those classified as having separable preferences, respondents with different unconditional preferences count north of 20%. A large measure of the effect is owed to the authors' inclusion of middle categories in the preference measures, and to the respondents' choosing DD or RR given the conditioning but a neutral position in the absence of it. However, there is a nontrivial amount of outright party switching (conditional to unconditional) in these data: 8.8% of those conditioned to DD chose R absent the conditioning; 4.7% conditioned to RR choose D. Using party identification as the unconditional measure, nearly thirty percent of voters classified as having separable preferences give a response different from the consistent ones they give on the conditional indicators. As with the other measurement standard, the partisanship version of unconditional party preference shows that the prevalent quirk in the data is the tendency for DD voters to chose a more Republican option without the conditioning. Indeed, the proportions of DD voters who seem more Republican in the unconditional measures are statistically distinct from the proportions of RR voters who seem unconditionally more Democratic ($P = .011$ using the "direct" measure of unconditional preference, $P = .016$ using party identification). If this is measurement error, it does not appear to be random, as we would expect these proportions to be indistinguishable from each other given the symmetry of the problem's context.

3   A Pseudo-classical Model of Voter Preference

The balancing problem is ripe for generalization. In doing so, however, we do not wish to fully foreclose on the simple model of §2.1 and its implications, first because it is clear from the data we and others have examined that this model ably characterizes the preferences of many voters, and that its distinctions between unifying and dividing have important empirical implications for statistical models of voter choice. Second, because an on-going interest in our research program is to better understand and articulate translations of models that have to-date been viewed in political science from strictly classical perspectives, we wish here to keep the one in §2.1 prominent in the background for the purpose of illuminating, if here in only a preliminary way, what kinships we can divine between the classical treatment of the problem and prominent advances in the QI literature. We begin by noting that there is no dispute in the U.S.-based political science literature over the dimensionality of the balancing problem, or others like it; indeed, in every paper we have seen that broaches the topic of separable versus nonseparable preferences, a simple example of what can be called the two-bit case is referenced in the course of explaining the distinction. However, as Smith et. al note, the weighted Euclidean has implications for the metrics of the space. The authors trace one implication into an analysis of the fit of statistical models, but as with others writing before and after them, they otherwise treat the weighted Euclidean as a tool to classify voters against hypothetical arrangements of parties and institutions – these voters evidencing separable preferences, these showing nonseparable preferences of a particular form, and so on. They do not, and indeed no classical scientist has, considered it as a cognitive process model, descriptive of the thinking of an individual voter.

When we do so, we see an important change in the dimensionality of the space when the elements of $\mathbf{A}$ are nonzero. Indeed, when as in §2.1 $\mathbf{A}$ is a matrix of ones, we see the problem applied to the voter as producing four and not two bits. This is because there are at the same time for every voter two partisan preferences for each dimension of governance, one (say, for the Legislature) "invoked" by a fixed (say, R) outcome in the other (Executive), and a second one, also for the legislature, "invoked" by the opposite fixed outcome ($D =$ Executive). So, for every voter, L can be conceptualized as a two-bit registry, and likewise for E, making the total four. Conceptually, we are now only a step away from a full alignment of the problem in a more general space, and indeed are outside the bounds of classical approaches already in considering two partisan preferences for each dimension at the same time for one voter. From a quantum perspective, such preferences would be described as being in superposition, and in a Hilbert space, the problem we have elaborated here would be described not in terms of four bits but rather in terms of two qubits. Such a space would generally have dimension 2^k, where k is the number of qubits. Defined over the real numbers, this space has deep kinships with spatial models in political science, including its depiction of distance, which is Euclidean. Defined over complex numbers, the problem is made still more general, as we would then have voters consider executive (e) and legislative (l) dimensions to governance and define partisan options (r and d) for each, writing the tensor product:

$$|e\rangle \otimes |l\rangle = a_0 b_0 |dd\rangle + a_1 b_0 |dr\rangle + a_0 b_1 |rd\rangle + a_1 b_1 |rr\rangle \tag{7}$$

where the two qubits are in superposition, e and l are independent, and the existence of the full complement of product weights defines separability.

In §2.1 we wrote of differentiating the square of the weighted Euclidean and finding four partial derivatives to invoke completely nonseparable preferences, one set shifting voters to the balancing regime and the other shifting other voters to the unifying regime. A quantum analogy of the four outcomes is:

$$\frac{1}{\sqrt{2}}(|dd\rangle + |rr\rangle) \qquad \frac{1}{\sqrt{2}}(|dd\rangle - |rr\rangle)$$

$$\frac{1}{\sqrt{2}}(|dr\rangle + |rd\rangle) \qquad \frac{1}{\sqrt{2}}(|dr\rangle - |rd\rangle)$$

These are the maximally-entangled, two-qubit states namesaked for John Bell after his fundamental work on the Einstein-Podolsky-Rosen paradox. In the QI literature, this type of nonseparability has been considered in application to social/cognitive data for more than a decade [16-20], and is of a radically different nature than any corollary ever considered in political science. Formally, "there are no coefficients which can decompose" [18] the states into the tensor product above that sets out e and l as independent. When conceptualized as resident in a Bell state, quantum nonseparable, preferences in one dimension are not so much conditioned or dependent upon outcomes in another as given by them, so even to properly consider those in one dimension requires consideration of those in the other.

Likewise, quantum separability differs radically from separability conditions customarily considered in political science. As fashioned above, the tensor product $|e\rangle \otimes |l\rangle$ seems to capture quite precisely the language used by political scientists when describing separability, as considerations in one dimension are independent of those in the other at the level of the voter. However, we have come to recognize that separable preferences as defined in the political science literature would not necessarily be viewed as separable in the quantum generalization. This can be seen in the tradition of preference order rankings understanding of separability by noting that one could not, against the quantum definition of separable preferences, write out a subset of the 4! preference orders in the problem and, a priori, privilege eight or four or even one of them as demonstrating separability. Likewise, classical reasoning from the spatial model and the weighted Euclidean about separability runs aground against the mathematics in (7). Indeed, the weighted norm fashioned so as to depict fully separable preferences they are traditionally understood in political science can be readily interpreted as mapping directly to one of the Bell states. The traditions, quantum versus classical, thus seem at once deeply related, and profoundly incompatible.

However, a recent and we think quite important innovation by Bruza, Iqbal, and Kitto [18] provides a passage into the boundary between Bell-type entanglement and nonseparability as it has been traditionally understood in political science. Challenging what perhaps was a status quo in the QI literature – using the Bell inequalities as "the formal device for determining non-separability" (p. 26) – the authors add to the nomenclature the notion of "psuedo-classical" nonseparability and situate understanding of this phenomenon in territory familiar to classically-trained social scientists by formalizing it as a factorization problem in a joint probability distribution. Probabilistically, if generalizing balancing theory fully quantum, we would refashion the state vector ψ from §2.1 to an uninformed state (ψ_u):

$$\psi = \begin{bmatrix} \psi DD \\ \psi DR \\ \psi RD \\ \psi RR \end{bmatrix} \longrightarrow \psi_u = \frac{1}{2} \begin{bmatrix} 1 \\ 1 \\ 1 \\ 1 \end{bmatrix} \tag{8}$$

where probabilistic reasoning must shift from within the confines of classical, Kolmogovorian theory to that of the more general theory often namesaked for Born [21].

The Bruza et al. [18] innovation in contrast foots the distinction between separable nonseparable in classical probability theory, retaining the law of total probability that is not a feature of the fully quantum perspective. Elaborating from a theorem proved by Suppes and Zanotti [22], [18] note that for two random variables A and B and a conditioning (factorizing) variable λ,

$$\Pr(A, B, \lambda) = \Pr(A|\lambda) \Pr(B|\lambda) \Pr(\lambda) \tag{9}$$

and

$$\Pr(A, B) = \sum_{j \in \Lambda} \Pr(A|\lambda_j) \Pr(B|\lambda_j) \Pr(\lambda_j) \tag{10}$$

where Λ is the set of values taken on by λ. In this framework we find a new lever into the balancing problem and the data in [6] by considering our executive party priming variable λ, and the respondents' choices over partisan Congressional control our central variable of interest Y. In analogous fashion to (10), we can write

$$\Pr(Y) = \sum_{j \in \Lambda} \Pr(Y|\lambda_j) \Pr(\lambda_j). \tag{11}$$

By treating the marginals of our unconditional Congressional control measure as an empirical estimate of the "true" unconditional distribution, we can compare (via a standard chi-square test) the cell frequencies for the two conditional measures to that for the unconditional item.

Bruza et al. note that, in addition to the law of total probability, their approach requires attention to the presumed prior probability distribution of λ, in particular that the distribution of λ is uniform. While in their experiments they randomly assigned subjects to priming conditions, here all respondents answer all three versions of the Congressional control question (conditional on Democratic control of the executive, conditional on Republican control of the executive, and unconditional). As a result, to ensure the robustness of our findings we consider the range of possible values for the prior on $\lambda = \Pr(\text{Dem. Executive Control})$; consistency in the findings of the test across a broad range of potential prior values for λ would suggest that our results are not sensitive to the choice of prior.

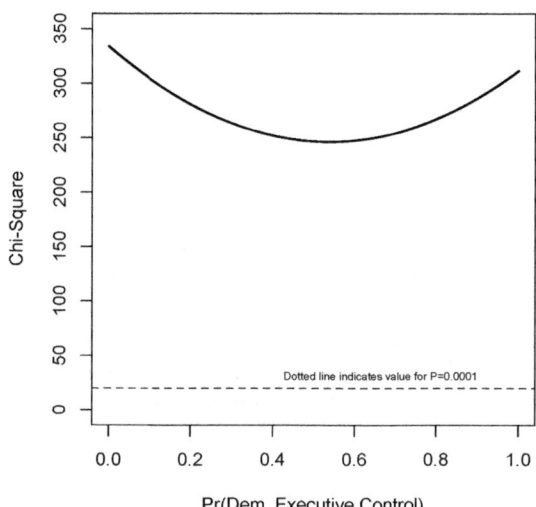

Fig. 2. χ^2 Values for Conditional vs. Unconditional Preferences Over Divided Government

Figure 2 plots the values of that χ^2 statistic over a range of values $\tilde{\lambda} \in [0, 1]$; cell frequencies for the statistic were thus calculated as $\tilde{\lambda} f_{Dj} + (1 - \tilde{\lambda} f_{Rj})$, where f_{Dj} and f_{Rj} denote the cell frequencies from the one-way table of responses conditional on Democratic and Republican control of the executive, respectively. For all possible prior values of λ, we note a substantial statistical difference between the distributions of preferences over partisan control of Congress between the conditional and unconditional measures, and at no point over the range of those values does the test statistic remotely approach statistical insignificance. As noted in [19], if the remaining two assumptions about the prior distribution of λ and the law of total probability hold, this can be interpreted as evidence in support of nonseparability in preferences.

In a recent paper, Busemeyer et al. [22] note that "quantum information processing principles provide a viable and promising new way to understand human judgment and reasoning." Somewhat more specifically, Bruza and colleagues suggest that their notion of "psuedo-classical" nonseparability "is a useful one in order to classify quantum-like systems" [18]. We are fully aligned with both of these these sentiments, and indeed have come to suspect that the latter will receive nontrivial attention and be considered as an outright alternative to what we have now come to view as very restrictive accounts in political science.

References

1. Fiorina, M.P.: Divided Government, 2nd edn. MacMillan, New York (1996)
2. Downs, A.: An Economic Theory of Democracy. Harper and Row, New York (1957)
3. Converse, P.E.: The Nature of Belief Systems in Mass Publics. In: Apter, D. (ed.) Ideology and Discontent. Free Press, New York (1964)
4. Petrocik, J., Doherty, J.: The Road to Divided Government: Paved without Intention. In: Galderisi, P.F., Herzberg, R.Q., McNamara, P. (eds.) Divided Government: Change, Uncertainty, and the Constitutional Order. Rowman & Littlefield, Lanham (1996)
5. Carsey, T., Layman, G.: Policy Balancing and Preferences for Party Control of Government. Political Research Quarterly 57, 541–550 (2004)
6. Smith Jr., C.E., Brown, R.D., Bruce, J.M., Marvin Overby, L.: Party Balancing and Voting for Congress in the 1996 National Election. American Journal of Political Science 43, 737–764 (1999)
7. Dean, L., Niou, E.M.S.: Elections in Double-Member Districts with Nonseparable Preferences. Journal of Theoretical Politics 10, 89–110 (1998)
8. Enelow, J.M., Hinich, M.J.: The Spatial Theory of Voting: An Introduction. Cambridge University Press, New York (1984)
9. Hinich, M.J., Munger, M.C.: Analytical Politics. Cambridge University Press, New York (1997)
10. Davis, O., Hinich, M.: On the Power and Importance of the Mean Preference in a Mathematical Model of Democratic Choice. Public Choice 5, 59–72 (1968)
11. Gorman, W.: The Structure of Utility Functions. Review of Economic Studies 32, 369–390 (1968)
12. Schwartz, T.: Collective Choice, Separation of Issues and Vote Trading. American Political Science Review 71, 999–1010 (1977)
13. Lacy, D.: A Theory of Nonseparable Preferences in Survey Responses. American Journal of Political Science 45, 239–258 (2001a)

14. Lacy, D.: Nonseparable Preferences, Measurement Error, and Unstable Survey Responses. Political Analysis 9, 95–115 (2001b)
15. Campbell, A., Converse, P.E., Miller, W., Stokes, D.: The American Voter. Wiley, New York (1960)
16. Aerts, D., Aerts, S., Broeckaert, J., Gabora, L.: The Violation of Bell Inequalities in the Macroworld. Foundations of Physics 30, 1378–1414 (2000)
17. Aerts, D., Gabora, L.: A Theory of Concepts and Their Combinations II: A Hilbert Space Representation. Kybernetes 34, 176–205 (2005)
18. Bruza, P., Iqbal, A., Kitto, K.: The Role of Non-Factorizability in Determining 'Pseudo-Classical' Non-Separability. In: Quantum Informantics for Cognitive, Social, and Semantic Processes: Papers from the AAAI Fall Symposium (2010), pp. 26–31 (2010)
19. Gabora, L., Aerts, D.: A Model of the Emergence and Evolution of Integrated Worldviews. Journal of Mathematical Psychology 53, 434–451 (2009)
20. Bruza, P.D., Kitto, K., Ramm, B., Sitbon, L., Song, D., Blomberg, S.: Quantum-like Non-Separability of Concept Combinations, Emergent Associates and Abduction. Logic Journal of the IGPL (2011) (forthcoming)
21. Born, M.: Zur Quantenmechanik der Stoßvorgänge. Zeitschrift für Physik 37, 863–867 (1926)
22. Busemeyer, J., Pothos, E., Franco, R., Trueblood, J.: A Quantum Theoretical Explanation for Probability Judgment 'Errors'. Psychological Review (2011) (forthcoming)
23. Trueblood, J., Busemeyer, J.: A Quantum Probability Explanation for Order Effects on Inference. Cognitive Science (2011) (forthcoming)
24. Weisberg, H.: A Multidimensional Conception of Party Identification. Political Behavior 2, 33–60 (1980)

A Quantum Cognition Analysis of the Ellsberg Paradox

Diederik Aerts, Bart D'Hooghe, and Sandro Sozzo

Center Leo Apostel, Brussels Free University
Krijgskundestraat 33, B-1160 Brussels, Belgium
{diraerts,bdhooghe,ssozzo}@vub.ac.be

Abstract. The *expected utility hypothesis* is one of the foundations of classical approaches to economics and decision theory and Savage's *Sure-Thing Principle* is a fundamental element of it. It has been put forward that real-life situations exist, illustrated by the *Allais* and *Ellsberg paradoxes*, in which the Sure-Thing Principle is violated, and where also the expected utility hypothesis does not hold. We have recently presented strong arguments for the presence of a double layer structure, a *classical logical* and a *quantum conceptual*, in human thought and that the quantum conceptual mode is responsible of the above violation. We consider in this paper the Ellsberg paradox, perform an experiment with real test subjects on the situation considered by Ellsberg, and use the collected data to elaborate a model for the conceptual landscape surrounding the decision situation of the paradox. We show that it is the overall conceptual landscape which gives rise to a violation of the Sure-Thing Principle and leads to the paradoxical situation discovered by Ellsberg.

Keywords: Sure-Thing Principle; Ellsberg paradox; conceptual landscape; quantum cognition.

1 Introduction

In game theory, decision theory and economics the *expected utility hypothesis* requires that individuals evaluate uncertain prospects according to their expected level of 'satisfaction' or 'utility'. In particular, the expected utility hypothesis is the predominant descriptive and normative model of choice under uncertainty in economics. From a mathematical point of view the expected utility hypothesis is founded on the *von Neumann-Morgenstern utility theory* [1]. These authors provided a set of 'reasonable' axioms under which the expected utility hypothesis holds. One of the axioms proposed by von Neumann and Morgenstern is the *independence axiom* which is an expression of Savage's *Sure-Thing Principle* [2], the latter being one of the building blocks of classical approaches to economics. Examples exist in the literature which show an inconsistency with the predictions of the expected utility hypothesis, namely a violation of the Sure-Thing Principle. These deviations, often called paradoxes, were firstly revealed by Maurice Allais [3] and Daniel Ellsberg [4]. The Allais and Ellsberg paradoxes

D. Song et al. (Eds.): QI 2011, LNCS 7052, pp. 95–104, 2011.
© Springer-Verlag Berlin Heidelberg 2011

at first sight at least, indicate the existence of an *ambiguity aversion*, that is, individuals prefer 'sure choices' over 'choices that contain ambiguity'. Several attempts have been put forward to solve the drawbacks raised by the Allais and Ellsberg paradoxes but none of the arguments that have been proposed is, at the best of our knowledge, considered as conclusive.

The above problems are strongly connected with difficulties that afflict cognitive science, i.e. the concept combination problem (see, e.g., [5]) and the disjunction effect (see, e.g., [6]). It is indeed so that concepts combine in human minds such that deviations are found from a manner of combination compatible with classical set and probability theories. Analogously, subjects take decisions which seem to contradict classical logic and probability theory. Trying to cope with these difficulties one of the authors has proposed, together with some coworkers, a formalism (*SCoP formalism*) in which context plays a relevant role in both concept combinations and decision processes [7,8,9]. Moreover, this role is very similar to the role played by the (measurement) context on microscopic particles in quantum mechanics. Within the SCoP perspective models have been elaborated which use the mathematical formalism of quantum mechanics to describe both concept combinations and the disjunction effect, and which accord with the experimental data existing in the literature [10,11,12,13]. This analysis has allowed the authors to suggest the hypothesis that two structured and superposed layers can be identified in human thought: a *classical logical layer*, that can be modeled by using a classical Kolmogorovian probability framework, and a *quantum conceptual layer*, that can instead be modeled by using the probabilistic formalism of quantum mechanics. The thought process in the latter layer is given form under the influence of the totality of the surrounding conceptual landscape, hence context effects are fundamental in this layer. The relevance of the quantum conceptual layer in producing the disjunction effect will be discussed in a forthcoming paper [14]. In the present paper we instead focus on the Ellsberg paradox. More precisely, after introducing Savage's Sure-Thing Principle and its violation occurring in the Ellsberg paradox in Sec. 2, we provide in Sec. 3 a preliminary analysis of the paradox, clarifying and fixing, in particular, some assumptions that are not made explicit in the standard presentations of it. Then, we discuss in Sec. 4 a real experiment on 59 test subjects that we have performed to test the Ellsberg paradox, and examine the obtained results. More specifically, we identify from the obtained answers and explanations the conceptual landscapes that we consider relevant in formulating the paradox. We finally work out in Sec. 5 the mathematical scheme for a quantum model in which each conceptual landscape is represented by a vector of a Hilbert space and the qualitative results obtained in our experiment are recovered by considering the overall conceptual landscape as the superposition of the single landscapes.

We conclude this section with some remarks. Firstly, we note that in our approach the explanation of the violation of the expected utility hypothesis and the Sure-Thing Principle is not (only) the presence of an ambiguity aversion. On the contrary, we argue that the above violation is due to the concurrence of superposed conceptual landscapes in human minds, of which some might be linked to ambiguity aversion, but other completely not. We therefore maintain that the violation of the Sure-Thing Principle should not be considered as a fallacy of human thought, as often claimed in the literature but, rather, as the proof that real subjects follow a different way of thinking than the one dictated by classical logic in some specific situations, which is context-dependent. Secondly, we observe that an explanation of the violation of the expected utility hypothesis and the Sure-Thing Principle in terms of quantum probability has already been presented in the literature (see, e.g., [15,16,17,18]). What is new in our approach is the fact that the quantum mechanical modeling is not just an elegant formal tool but, rather, it reveals an underlying quantum conceptual thought. Thirdly, the presence of a quantum structure in cognition and decision making does not necessarily presuppose the existence of microscopic quantum processes in human brain. Indeed, we avoid such a compelling assumption in our approach.

2   The Sure-Thing Principle and the Ellsberg Paradox

Savage introduced the *Sure-Thing Principle* [2] inspired by the following story.

A businessman contemplates buying a certain piece of property. He considers the outcome of the next presidential election relevant. So, to clarify the matter to himself, he asks whether he would buy if he knew that the Democratic candidate were going to win, and decides that he would. Similarly, he considers whether he would buy if he knew that the Republican candidate were going to win, and again finds that he would. Seeing that he would buy in either event, he decides that he should buy, even though he does not know which event obtains, or will obtain, as we would ordinarily say.

The Sure-Thing Principle is equivalent to the independence axiom of expected utility theory: 'independence' here means that if persons are indifferent in choosing between simple lotteries L_1 and L_2, they will also be indifferent in choosing between L_1 mixed with an arbitrary simple lottery L_3 with probability p and L_2 mixed with L_3 with the same probability p.

Let us consider the situation put forward by Daniel Ellsberg [4] to point out an inconsistency with the predictions of the expected utility hypothesis and a violation of the Sure-Thing Principle. Consider an urn known to contain 30 red balls and 60 balls that are either black or yellow, the latter in unknown proportion. One ball is to be drawn at random from the urn. To 'bet on red' means that you will receive a prize a (say, 10 euros) if you draw a red ball ('if red occurs') and a smaller amount b (say, 0 euros) if you do not. If test subjects are given the following 4 options: (I) 'a bet on red', (II) 'a bet on black', (III) 'a bet on red or yellow', (IV) 'a bet on black or yellow', and are then presented with the choice between bet I and bet II, and the choice between bet III and bet IV, it appears that a very frequent pattern of response is that bet I is preferred to bet II, and bet IV is preferred to bet III. This violates the Sure-Thing Principle, which requires the ordering of I to II to be preserved in III and IV (since these two pairs differ only in the pay-off when a yellow ball is drawn, which is constant for each pair). The first pattern, for example, implies that test subjects bet on (against) red rather than on (against) black.

The contradiction above suggests that preferences of 'real-life' subjects are inconsistent with Savage's Sure-Thing Principle of expected utility theory. A possible explanation of this drawback could be that people make a mistake in their choice and that the paradox is caused by an error of reasoning. In our view, however, these examples show that subjects make their decisions in ways which do violate the Sure-Thing Principle, but not because they make an error of reasoning. Rather, this occurs because they follow a different type of reasoning which is not only guided by logic but also by conceptual thinking which is structurally related to quantum mechanics. We stress that in the Ellsberg paradox the situation where the number of yellow balls and the number of black balls are not known individually, only their sum being known to be 60, introduces the so-called *disjunction effect* [6], which is systematically discussed in [14].

3   A Preliminary Analysis of the Paradox

Frank Knight introduced a distinction between different types of uncertainty [19], and Daniel Ellsberg stimulated the reflection about them [4]. More explicitly, Ellsberg put forward the notion of *ambiguity* as an uncertainty without any well-defined probability measure to model this uncertainty, as opposed to *risk*, where such a probability measure does exist. In the case of the Ellsberg paradox situation, 'betting on red' concerns a situation in which the uncertainty is modeled by a probability measure which is given, namely a probability of $\frac{1}{3}$ to win the bet, and a probability of $\frac{2}{3}$ to lose it. For 'betting on black', however, the situation is such that no definite probability measure models the situation related to this bet. Indeed, since it is only known that the sum of the black and the yellow balls is 60, the number of black balls is not known. If no additional information is given specifying in more detail the situation of the Ellsberg paradox, 'betting on black' will be a situation of ambiguity, since the probability measure associated with this bet is not known. Of course, by making a specific additional assumption, namely the assumption that black and yellow balls are chosen at random until their sum reaches 60, we can re-introduce a probability measure corresponding to the 'bet on black' situation. In this case, also for 'betting on black' the probability of winning equals $\frac{1}{3}$ and that of losing equals $\frac{2}{3}$. If the Ellsberg paradox situation is presented as a real-life situation, for reasons of symmetry, it can be supposed that indeed black and yellow balls are chosen at random until their sum reaches 60, and then put in the urn. In this case a 'bet on black' is equivalent with a 'bet on red'.

However, there are many possible situations of 'real life' where this symmetry is perhaps not present, one obvious example being the one where the person proposing to bet following an Ellsberg type of situation has the intention to trick, and for example installs a way to have systematically less black balls than yellow balls in the urn. Of course, the real aim of the Ellsberg paradox is to show that 'people will already take into account this possibility' even if nothing is mentioned extra, which means that most probably the situation is symmetric. We will see that our analysis by means of the introduction of different conceptual landscapes sheds light on this aspect of the paradox.

In the following we analyze the Ellsberg paradox situation, using the explanation we introduced for the presence of underextension and overextension for concept combinations and for the disjunction effect [10,11]. The essential element of our explanation is the distinction between 'the conceptual landscape surrounding a given situation' and the 'physical reality of this given situation'. The probabilities governing human decisions are related to this conceptual landscape and not necessarily to the physical reality of a given situation. Although there is a correspondence between the physical reality of a situation and the surrounding conceptual landscape, in most cases this correspondence is far from being an isomorphism. For the situation of the Ellsberg paradox, let us first describe the physical reality of the situation and then provide a plausible conceptual landscape surrounding this situation.

The physical situation is the urn containing red, black and yellow balls, with the number of red balls being 30 and the sum of the number of black balls and yellow balls being 60. The original article [4] does not specify the physical situation in any further detail, leaving open the question as to 'how the black and the yellow balls are chosen when 60 of them are put in the urn'. We prefer to make the physical situation more specific and introduce an additional hypothesis, namely that the black and the yellow balls are put in the urn according to a coin toss. When heads is up, a black ball is added to the urn, and when tails is up a yellow ball is added. Prepared according to the Ellsberg situation, the urn will contain 30 red balls, $60 - n$ black balls and n yellow balls, where $n \in \{0, 1, \ldots, 59, 60\}$. In this case, when we choose a ball at random, there is a probability of $\frac{1}{3}$ for a red ball to turn up, a probability of $\frac{60-n}{90}$ for a black ball to turn up, and a probability of $\frac{n}{90}$ for a yellow ball to turn up. For an urn prepared according to the outcome of a coin toss, however, the probability for red to turn up is $\frac{1}{3}$, the probability for black to turn up is $\frac{1}{3}$, and the probability for yellow to turn up is also $\frac{1}{3}$.

4   An Experiment Testing the Ellsberg Paradox

For the type of analysis we make, we need to account for different pieces of conceptual landscape. To gather relevant information, we decided to perform a test of the Ellsberg paradox problem. Thus, we sent out the following text to several friends, relatives and students. We asked them to forward our request to others, so that our list could also include people we didn't know personally.

We are conducting a small-scale statistics investigation into a particular problem and would like to invite you to participate as test subjects. Please note that it is not the aim for this problem to be resolved in terms of correct or incorrect answers. It is your preference for a particular choice we want to test. The question concerns the following situation.

Imagine an urn containing 90 balls of three different colors: red balls, black balls and yellow balls. We know that the number of red balls is 30 and that the sum of the black balls and the yellow balls is 60. The questions of our investigation are about the situation where somebody randomly takes one ball from the urn.

- The first question is about a choice to be made between two bets: bet I and bet II. Bet I involves winning '10 euros when the ball is red' and 'zero euros when it is black or yellow'. Bet II involves winning '10 euros when the ball is black' and 'zero euros when it is red or yellow'. The first question we would ask you to answer is: Which of the two bets, bet I or bet II, would you prefer?

- The second question is again about a choice between two different bets, bet III and bet IV. Bet III involves winning '10 euros when the ball is red or yellow' and 'zero euros when the ball is black'. Bet IV involves winning '10 euros when the ball is black or yellow' and 'zero euros when the ball is red'. The second question therefore is: Which of the two bets, bet III or bet IV, would you prefer?

Please provide in your reply message the following information:

For question 1, your preference (your choice between bet I and bet II). For question 2, your preference (your choice between bet III and bet IV).

By 'preference' we mean 'the bet you would take if this situation happened to you in real life'. You are expected to choose one of the bets for each of the questions, i.e. 'not choosing is no option'.

You are welcome to provide a brief explanation of your preferences, which may be of a purely intuitive nature, only mentioning feelings, for example, but this is not required. It is allright if you only state your preferences without giving any explanation.

One final remark about the colors. Your choices should not be affected by any personal color preference. If you feel that the colors of the example somehow have an influence on your choices, you should restate the problem and take colors that are indifferent to you or, if this does not work, other neutral characteristics to distinguish the balls.

Let us now analyze the obtained results.

We had 59 respondents participating in our test of the Ellsberg paradox problem, of whom 34 preferred bets I and IV, 12 preferred bets II and III, 7 preferred bets II and IV and 6 preferred bets I and III. This makes the weights with preference of bet I over bet II to be 0.68 against 0.32, and the weights with preference of bet IV over bet III to be 0.71 against 0.29. It is interesting to note that 34+12=46 people chose the combination of bet I and bet IV or bet II and bet III, which is 78%. Of the 59 participants there were 10 who provided us an explanation for their choice. Interestingly, an independent consideration of this group of 10 reveals a substantial deviation of their statistics from the overall statistics: only 4 of them chose bet I and bet IV, 2 chose bet II and bet III, 3 chose bet II and bet IV, and 1 chose bet I and bet III. What is even more interesting, however, is that only half of them preferred bet I to bet II. So the participants in the 'explaining sub-group' were as likely to choose bet I as they were likely to choose bet II. This is too small a sample of 'subjects providing an explanation' to be able to make a firm conclusion about the different pieces of conceptual landscape in this Ellsberg paradox situation. Since this article is mainly intended to illustrate our way of modeling the situation, we will make a proposal for such a possible conceptual landscape.

A first piece of conceptual landscape is: 'an urn is filled with 30 balls that are red, and with 60 balls chosen at random from a collection of black and a collection of yellow balls'. We call this piece of conceptual landscape the *Physical Landscape*. It represents that which is most likely to correspond to the physical presence of an actual Ellsberg paradox situation. A second piece of conceptual landscape is: 'there might well be substantially fewer black balls than yellow balls in the urn, and so also substantially fewer black balls than red balls'. We call this piece of conceptual landscape the *First Choice Pessimistic Landscape*. It represents a guess of a less advantageous situation compared to the neutral physical one, when the subject is reflecting on the first choice to be made. A third piece of conceptual landscape is: 'there might well be substantially more black balls than yellow balls in the urn, and so also substantially more black balls than red balls'. This third piece we call the *First Choice Optimistic Landscape*. It represents a guess of a more advantage situation compared to the neutral physical one, when the subject is reflecting on the first choice to be made. A fourth piece of conceptual landscape is: 'there might well be substantially fewer yellow balls than black balls, and so substantially fewer red plus yellow balls than black plus yellow balls, of which there are a fixed number, namely 60'. This fourth piece we call *Second Choice Pessimistic Landscape*. It represents a guess of a less advantageous situation compared to the neutral physical one, when the subject is reflecting on the second choice to be made. A fifth piece of conceptual landscape is: 'there might well be substantially more yellow balls than black balls, and so substantially more red plus yellow balls than black plus yellow balls, of which there are a fixed number, namely 60'. This fifth piece we call the *Second Choice Optimistic Landscape*. It represents a guess of a more advantageous situation compared to the neutral physical one, when the subject is reflecting on the second choice to be made. A sixth piece of conceptual landscape, which we call the *Suspicion Landscape*, is: 'who knows how well the urn has been prepared, because after all, to put in 30 red balls is straightforward enough, but to pick 60 black and yellow balls is quite another thing; who knows whether this is a fair selection or somehow a biased operation, there may even have been some kind of trickery involved'. A seventh piece of conceptual landscape is: 'if things become too complicated I'll bet on the simple situation, the situation I understand well', which we call the *Don't Bother Me With Complications Landscape*.

These pieces of conceptual landscape are the ones we can reconstruct taking into account the explanations we received from our test subjects. We are convinced, however, that they are by no means the only possible relevant pieces of conceptual landscape. For example, one of the subjects who participated in our test and chose bet II and bet III said that she would have chosen differently, preferring bet I and bet IV, if more money had been involved. This leads us to believe that what plays a major role too in the choices the subjects make is whether they regard the test as a kind of funny game or make a genuine attempt to try and guess what they would do in real life when presented with a betting situation of the Ellsberg type. At an even more subtle level, subjects who feel that by choosing the combination bet I and bet IV, they would be choosing for a greater degree of predictability, might be tempted to change their choice, preferring the more unpredictable combination of bet II and bet III, because this is intellectually more challenging, although again this would depend on how they conceive the situation. Indeed, we firmly believe that the determining of further conceptual landscapes that are relevant involves even more subtle aspects.

5   A Quantum Model for Conceptual Landscapes

Let us illustrate in this section how a quantum modeling scheme can be worked out using the conceptual landscapes introduced in Sec. 4.

Consider the piece of conceptual landscape which we called the *Physical Landscape*, and suppose that it is the only piece, i.e. that it constitutes the whole conceptual landscape for a specific individual subject. This means this subject has no preference for bet I or bet II, and also has no preference for bet III or bet IV, so that the Sure-Thing Principle is not violated. A simple quantum mechanical model of this situation is one where we represent the conceptual landscape by means of vector $|A\rangle$, and the choice between bet I and bet II by means of a projection operator M such that $\mu_M(A) = \langle A|M|A\rangle$ is the weight for a subject to choose bet I, while $1 - \mu_M(A) = \langle A|1 - M|A\rangle$ is the weight for a subject to choose bet II, while the choice between bet III and bet IV is described by a projection operator N such that $\mu_N(A) = \langle A|N|A\rangle$ is the weight for a subject to choose bet III, while $1 - \mu_N(A) = \langle A|1 - N|A\rangle$ is the weight for a subject to choose bet IV. We have $\mu_M(A) = \mu_N(A) = \frac{1}{2}$.

Consider now the piece of conceptual landscape *First Choice Pessimistic*, and suppose that this is the only piece of conceptual landscape. Then bet I will be strongly preferred over bet II, and a quantum modeling of this situation consists in representing this piece of conceptual landscape by means of a vector $|B\rangle$ such that $\mu_M(B) = \langle B|M|B\rangle$ and $1 - \mu_M(B) = \langle B|1 - M|B\rangle$ represent the weights for subjects to choose bet I and bet II, respectively, so that $1 - \mu_M(B) \ll \mu_M(B)$ or, equivalently, $\frac{1}{2} \ll \mu_M(B)$. It is not easy to know how $\mu_N(B)$ will be under conceptual landscape *First Choice Pessimistic*. Indeed, our experience with the test we conducted indicates that, when subjects are asked to compare bet III and bet IV, other conceptual landscapes become relevant and predominant than the conceptual landscapes that are relevant and predominant when they are asked to compare bet I and bet II. Subjects who tend to give a high weight to conceptual landscape *First Choice Pessimistic* when comparing bet I and bet II, i.e. 'who fear that there might be substantially fewer black balls than red balls' seem to focus rather on the variability of the yellow balls when asked to compare bet III and bet IV, and tend to give dominance to conceptual landscape *Second Choice Pessimistic*, 'fearing that there might be substantially fewer yellow balls than black balls, and hence also fewer red plus yellow balls than black plus yellow balls'. This is borne out by the fact that 46 people, or 78% of the total number of participants, choose for the combination of bet *I* and bet *IV* or bet *II* and bet *III*. However, we also noted that some subjects gave dominance to what we have called conceptual landscape *Don't Bother Me With Complications* when they were asked to choose between bet III and bet IV. They had preferred bet I to bet II, and now also preferred bet III to bet IV. When asked why they preferred bet I to bet II, their answer was 'because we know what the risk is for red, but for black we do not'. Interestingly, when we asked them to reconsider their choice with respect to bet III and bet IV – they had preferred bet III – now explaining to them that bet IV gave rise to 'less uncertainty' than bet III, they remained with their preference for bet III to bet IV, commenting that 'anyhow betting on red made them feel more comfortable much like when asked to choose between bet I and bet II'. We believe that the rather artificial aspect of choosing between bet III and bet IV, of considering outcomes whose definitions are disjunctions of simple outcomes, makes this choice essentially more complicated, such that the choices made by these subjects are in line with what the Ellsberg paradox analysis tries to put forward. However, due to the relatively greater complexity of bet III and bet IV, as compared to bet I and bet II, this aspect is not revealed.

Anyhow, considerations like the one above are not our primary concern here, since we mainly want to give an account of how we apply our quantum-conceptual modeling scheme in the situation we have described. Again, because of the rather limited nature of the experiment conducted for this article, we have not been able to estimate the value of $\mu_N(B)$. However, if we call $|D\rangle$ the vector representing the conceptual landscape *Second Choice Pessimistic*, we have $\mu_N(D) \ll \frac{1}{2}$. If $|C\rangle$ and $|E\rangle$ represent the *First Choice Optimistic Landscape* and the *Second Choice Optimistic Landscape*, we have $\mu_M(C) \ll \frac{1}{2}$ and $\frac{1}{2} \ll \mu_N(E)$, respectively.

Let us now look at the *Suspicion Landscape* and represent it by the vector $|F\rangle$. In this case, we have $\frac{1}{2} \ll \mu_M(F)$ and $\mu_N(F) \ll \frac{1}{2}$, i.e. we have a situation that resembles what is generally claimed with respect to the Ellsberg paradox situation, which entails a violation of the Sure-Thing Principle. Finally, let us represent the *Don't Bother Me With Complications Landscape* by the vector $|G\rangle$. Then, $\frac{1}{2} \ll \mu_M(G)$ and $\frac{1}{2} \ll \mu_N(G)$, which instead does not violate the Sure-Thing Principle. Following the general quantum modeling scheme we worked out in detail in earlier publications [7,8,9,10,11,12,13], when all these pieces of conceptual landscape are present with different weights, the vector to model this situation is a normalized superposition of the vectors $|A\rangle, |B\rangle, |C\rangle, |D\rangle, |E\rangle, |F\rangle$ and $|G\rangle$. This makes it possible to choose coefficients of superposition such that if the Ellsberg paradox situation is surrounded by the conjunction of all these pieces of conceptual landscape, the Sure-Thing Principle will be violated in a way corresponding to experimental data that are collected with respect to this situation.

To conclude, we have recently introduced a notion of *contextual risk* to model the context dependent situations that are described in the literature in terms of ambiguity. Then, we have employed our *hidden measurement formalism* to show that these situations must be probabilistically described in a non-Kolmogorovian quantum-like framework [20], and we have provided a sphere model for the Ellsberg paradox [21], thus providing a concrete support to the employment of quantum-like structures in these situations.

Acknowledgments. The authors are greatly indebted with the 59 friends and colleagues for participating in the experiment. This research was supported by Grants G.0405.08 and G.0234.08 of the Flemish Fund for Scientific Research.

References

1. von Neumann, J., Morgenstern, O.: Theory of Games and Economic Behavior. Princeton University Press, Princeton (1944)
2. Savage, L.J.: The Foundations of Statistics. Wiley, New York (1954)
3. Allais, M.: Le Comportement de l'Homme Rationnel Devant le Risque: Critique des Postulats et Axiomes de l'École Américaine. Econometrica 21, 503–546 (1953)
4. Ellsberg, D.: Risk, Ambiguity, and the Savage Axioms. Quart. J. Econ. 75(4), 643–669 (1961)
5. Hampton, J.A.: Disjunction of Natural Concepts. Memory & Cognition 16, 579–591 (1988)
6. Tversky, A., Shafir, E.: The Disjunction Effect in Choice Under Uncertainty. Psych. Sci. 3, 305–309 (1992)
7. Gabora, L., Aerts, D.: Contextualizing Concepts Using a Mathematical Generalization of the Quantum Formalism. J. Exp. Theor. Art. Int. 14, 327–358 (2002)
8. Aerts, D., Gabora, L.: A Theory of Concepts and Their Combinations I: The Structure of the Sets of Contexts and Properties. Kybernetes 34, 167–191 (2005)
9. Aerts, D., Gabora, L.: A Theory of Concepts and Their Combinations II: A Hilbert Space Representation. Kybernetes 34, 192–221 (2005)
10. Aerts, D.: Quantum Structure in Cognition. J. Math. Psych. 53, 314–348 (2009)
11. Aerts, D., D'Hooghe, B.: Classical Logical Versus Quantum Conceptual Thought: Examples in Economics, Decision Theory and Concept Theory. In: Bruza, P., Sofge, D., Lawless, W., van Rijsbergen, K., Klusch, M. (eds.) QI 2009. LNCS, vol. 5494, pp. 128–142. Springer, Heidelberg (2009)
12. Aerts, D.: Quantum Interference and Superposition in Cognition: Development of a Theory for the Disjunction of Concepts. In: Aerts, D., D'Hooghe, B., Note, N. (eds.) Worldviews, Science and Us: Bridging Knowledge and Its Implications for Our Perspectives of the World. World Scientific, Singapore (2011) (in print) Archive reference and link, http://arxiv.org/abs/0705.1740 (2007)
13. Aerts, D.: General Quantum Modeling of Combining Concepts: A Quantum Field Model in Fock Space (2007), Archive reference and link: http://arxiv.org/abs/0705.1740
14. Aerts, D., Broekaert, J., Czachor, M., D'Hooghe, B.: A Quantum-Conceptual Explanation of Violations of Expected Utility in Economics. In: Song, D., et al. (eds.) QI 2011. LNCS, vol. 7052, pp. 192–198. Springer, Heidelberg (2011)
15. Busemeyer, J.R., Wang, Z., Townsend, J.T.: Quantum Dynamics of Human Decision-Making. J. Math. Psych. 50, 220–241 (2006)
16. Franco, R.: Risk, Ambiguity and Quantum Decision Theory(2007), Archive reference and link: http://arxiv.org/abs/0711.0886
17. Khrennikov, A.Y., Haven, E.: Quantum Mechanics and Violations of the Sure-Thing Principle: The Use of Probability Interference and Other Concepts. J. Math. Psych. 53, 378–388 (2009)
18. Pothos, E.M., Busemeyer, J.R.: A Quantum Probability Explanation for Violations of 'Rational' Decision Theory. Proc. Roy. Soc. B 276, 2171–2178 (2009)
19. Knight, F.H.: Risk, Uncertainty and Profit. Houghton Mifflin, Boston (1921)
20. Aerts, D., Sozzo, S.: Contextual risk and its relevance in economics. Accepted for Publication in J. Eng. Sci. Tech. Rev. (2011)
21. Aerts, D., Sozzo, S.: A contextual risk model for the Ellsberg paradox. Accepted for Publication in J. Eng. Sci. Tech. Rev. (2011)

Can Classical Epistemic States Be Entangled?

Harald Atmanspacher[1,4,5], Peter beim Graben[2,6], and Thomas Filk[1,3,4]

[1] Institute for Frontier Areas of Psychology, Freiburg, Germany
[2] Department of Linguistics, Humboldt University, Berlin, Germany
[3] Institute of Physics, University of Freiburg, Germany
[4] Parmenides Center for the Study of Thinking, Munich, Germany
[5] Collegium Helveticum, ETH Zürich, Switzerland
[6] Bernstein Center for Computational Neuroscience, Berlin, Germany

Abstract. Entanglement is a well-known and central concept in quantum theory, where it expresses a fundamental nonlocality (holism) of ontic quantum states, regarded as independent of epistemic means of gathering knowledge about them. An alternative, epistemic kind of entanglement is proposed for epistemic states (distributions) of dynamical systems represented in classical phase spaces. We conjecture that epistemic entanglement is to be expected if the states are based on improper phase space partitions. The construction of proper partitions crucially depends on the system dynamics.

Although improper partitions have a number of undesirable consequences for the characterization of dynamical systems, they offer the potential to understand some interesting features such as incompatible descriptions, which are typical for complex systems. Epistemic entanglement due to improper partitions may give rise to epistemic classical states analogous to quantum superposition states. In mental systems, interesting candidates for such states have been coined acategorial states, and among their key features are temporally nonlocal correlations. These correlations can be related to the situation of epistemic entanglement.

Keywords: non-commuting operations, phase space partitions, dynamical entropy, incompatibility, symbolic dynamics, epistemic entanglement, acategorial mental states, temporal nonlocality.

1 Introduction

It has been an old idea by Niels Bohr that central conceptual features of quantum theory, such as complementarity, are also of pivotal significance far exceeding the domain of physics. Although Bohr was always convinced of the extraphysical relevance of complementarity, he never elaborated this idea in concrete detail, and for a long time after him no one else did so either.

By now, a number of research programs have been developed in order to pick up Bohr's proposal with particular respect to psychology and cognitive science. The first steps in this direction were made by the group of Aerts in the early 1990s (Aerts et al. 1993), using non-distributive propositional lattices to address quantum-like behavior in non-quantum systems. Alternative approaches

D. Song et al. (Eds.): QI 2011, LNCS 7052, pp. 105–115, 2011.

have been initiated by Khrennikov (1999), focusing on non-classical probabilities, and Atmanspacher et al. (2002), outlining an algebraic framework with non-commuting operations. Two other, more recent lines of thinking are due to Primas (2007), addressing complementarity with partial Boolean algebras, and Filk and von Müller (2008), indicating strong links between basic conceptual categories in quantum physics and psychology.

Intuitively, it is quite unproblematic to understand why non-commuting operations or non-Boolean logic should be relevant, even inevitable, for mental systems that have nothing to do with quantum physics. Simply speaking, the non-commutativity of operations means nothing else than that the sequence, in which operations are applied, matters for the final result. And non-Boolean logic refers to propositions that may have unsharp truth values beyond yes or no, shades of plausibility or credibility as it were. Both versions obviously abound in psychology and cognitive science (and in everyday life), and they have led to well-defined and specific theoretical models with empirical confirmation and novel predictions. Five kinds of psychological phenomena have been addressed so far: (i) decision processes, (ii) semantic networks, (iii) bistable perception, (iv) learning, and (v) order effects in questionnaires (see Atmanspacher 2011, Sec. 4.7, for a compact review).

In earlier publications (beim Graben and Atmanspacher 2006, 2009) we studied in detail how the concept of complementarity can be sensibly addressed in classical dynamical systems as represented in a suitable phase space. The formal key to such a generalized version of complementarity lies in the construction of phase space partitions, which give rise to epistemic states. Descriptions based on partitions are compatible only under very specific conditions, otherwise they are incompatible or complementary. In this paper we ask whether entanglement, another central feature of quantum theory, may also be given meaning in the same framework.

2   Non-commutative Operations

Non-commutative operations are at the core of quantum physics, where they appear as elements of algebras of observables. But non-commutative operations also abound in classical physical systems, as has been discussed frequently (see a recent paper by beim Graben and Atmanspacher (2006) including references given therein). A significant field in which this has become apparent is the theory of complex dynamical systems in physics.

Particularly fertile playgrounds for non-commutativity are complex systems outside physics for which interactions with their state (expressed as actions of an operator) are explicitly known to inevitably change that state. This is invariably the case in psychology: every interaction with a mental state changes that state in a way making it virtually impossible to prepare or re-prepare mental states strictly identically.

An intuitively appealing characterization of non-commutative operations A and B is to say that the sequence, written as multiplication, in which A and B

are applied to a state makes a difference:

$$AB \neq BA. \tag{1}$$

If an addition of operations is defined as well, one can write:

$$[A, B] = AB - BA \neq 0, \tag{2}$$

and, given a commutator C, we have:

$$[A, B] = AB - BA = C \tag{3}$$

In quantum physics, the commutator for canonically conjugate quantum observables is universal: $C = h \cdot \mathbb{1}$, with h as the Planck action. For complex physical systems, and even more so for mental systems, we can hardly expect the commutator to be universal, but we may hope to find regularities for equivalence classes of systems. At present, we do not know how to do this in a deductive theoretical fashion, but there is a possibility to approach the problem empirically.

Commutation relations between two non-commuting operations A, B generically entail an uncertainty relation

$$\Delta A \cdot \Delta B \geq 1/2 |\langle C \rangle|, \tag{4}$$

where ΔA and ΔB are the variances of measured distributions of A and B, and $\langle C \rangle$ is the expectation value of C. Changing the conditions under which A and B are measured, it should be possible to investigate how the variances covary, and thus (at least) to estimate a lower bound for $\langle C \rangle$.

For the representation of commutation relations, i.e. of the way in which operators act on states, we need to specify a representation space. While this is typically chosen as a Hilbert space in quantum physics, a preferrable option for classical systems is a symplectic phase space or, more generally for complex systems, even a phase space without symplectic structure. In this contribution, we refer to the notion of a phase space in this general sense.

3   Phase Space Partitions

In the theory of dynamical systems, the state of a system is usually represented by a subset of its phase space Ω. For classical systems, their *ontic state* at a given time t is represented by a point $x \in \Omega$, while an *epistemic state* can be represented as a region $A \in \Omega$ comprising many ontic states.[1] More formally speaking, epistemic states are subsets $A_1, A_2, ..., A_n$ of Ω with $A_i \cap A_j = \emptyset$ and $\bigcup_{i=1,...,n} A_i = \Omega$.

---

[1] More precisely, epistemic states are distributions in a probability space over Ω, but for the present discussion it is sufficient to consider their support $A \in \Omega$; see beim Graben and Atmanspacher (2006, 2009). For a detailed discussion of ontic and epistemic states see Primas (1990) and Atmanspacher and Primas (2003) or, as a related framework, Spekkens (2007) and Harrigan and Spekkens (2010).

If f is an observable of the system considered, then f ascribes a valuation to its states. For ontic states x, this valuation is simply $f(x)$, but for epistemic states A_i the situation is different: in the simplest case their valuation f is the same for all ontic states in the same subset A_i: $f(x) = f(y)$ for all ontic states $x, y \in A_i$. In this case, x and y are *epistemically equivalent* with respect to f.

The set $\mathcal{F} = \{A_1, A_2, ..., A_n\}$ of all subsets A_i is called a *phase space partition*.

- If every A_i is a singleton, i.e. represents an ontic state, $\mathcal{F}$ is the identity partition $\mathcal{I}$.
- If $A_1 = \Omega$, i.e. the entire phase space, $\mathcal{F}$ is the trivial partition.
- If $\mathcal{F}$ and $\mathcal{G}$ are finite partitions, $\mathcal{P} = \mathcal{F} \vee \mathcal{G} = \{A_i \cap B_j\}$ is a product partition.

4   Dynamics

Let us now consider the time evolution of the system, i.e. its dynamics, generated by a flow operator Φ acting on an ontic state x_t at time t,

$$x_{t+1} = \Phi^{t+1}(x_o) = \Phi(\Phi^t(x_o)) = \Phi(x_t), \tag{5}$$

and combine this dynamics with the action of an observable f. The valuation $f(x_o)$ applies to an ontic state x_o in the epistemic state $A_{i_o} \in \mathcal{F}$. Similarly, $f(x_1) = f(\Phi(x_o))$ applies to an ontic state x_1 in the epistemic state $A_{i_1} \in \mathcal{F}$. This way, measuring $f(x_1)$ yields information about x_o, namely that x_o is contained in the epistemic state given by the intersection of A_{i_o} with the pre-image of A_{i_1}, $A_{i_o} \cap \Phi^{-1}(A_{i_1})$. We can continue this procedure iteratively up to measurements of x_n and obtain the information which measuring $f(x_n)$ yields about all previous states $x_{i<n}$.

Rather than talking about pre-images Φ^{-t} of epistemic states A_i, we generalize the terminology and refer to pre-images of the partition as a whole, $\Phi^{-1}(\mathcal{F}) = \{\Phi^{-1}(A_i)\}$. This allows us to define the dynamic refinement of $\mathcal{F}$ as a product partition $\mathcal{F} \vee \Phi^{-1}(\mathcal{F})$. The finest refinement $R\mathcal{F}$ is obtained in the limit $t \to \pm\infty$:

$$R\mathcal{F} = \bigvee_{t=-\infty}^{\infty} \Phi^{-t}(\mathcal{F}) \tag{6}$$

If $R\mathcal{F} = \mathcal{I}$, the partition $\mathcal{F}$ is the *generating partition* $\mathcal{P}_g$. It is distinguished by the fact that measurements of f yield complete information about the ultimate pre-image x_o of all epistemic states and, thus, gives rise to the determination of x_o as a dispersion-free ontic state. If $R\mathcal{F} \neq \mathcal{I}$, no dynamic refinement leads to such dispersion-free states.

5   Dynamical Entropy

For a partition $\mathcal{F} = (A_1, A_2, ..., A_n)$ of a state space Ω, a simple version of the *entropy* of the system is the well-known Shannon entropy

$$H(\mathcal{F}) = -\sum_{i=1}^{n} \mu(A_i) \, \log \mu(A_i), \tag{7}$$

where $\mu(A_i)$ is the probability that the system state resides in partition cell A_i.

The *dynamical entropy* of a system in Ω requires us to consider its dynamics $\Phi : \Omega \to \Omega$ with respect to a partition $\mathcal{F}$:

$$H(\Phi, \mathcal{F}) = \lim_{n\to\infty} \frac{1}{n} H(\mathcal{F} \vee \Phi\mathcal{F} \vee \dots \vee \Phi^{n-1}\mathcal{F}) \tag{8}$$

In other words, the dynamical entropy is the limit of the Shannon entropy of the product partition of increasing dynamical refinement.

An important upper bound for the dynamical entropy is the *Kolmogorov-Sinai entropy* (Kolmogorov 1958, Sinai 1959). It is defined as the supremum of the dynamical entropy over all partitions $\mathcal{F}$,

$$H_{KS} = \sup_{\mathcal{F}} H(\Phi, \mathcal{F}), \tag{9}$$

and it is assumed if $\mathcal{F}$ is a *generating partition* $\mathcal{P}_g$, so that $H_{KS} = H(\Phi, \mathcal{P}_g)$. If $\mathcal{F}$ is not generating, $H(\Phi, \mathcal{F}) < H_{KS}$.

Maximizing the dynamical entropy, $\mathcal{P}_g$ minimizes correlations among partition cells such that only correlations due to the dynamics Φ itself contribute to $H(\Phi, \mathcal{P}_g)$. This can be understood due to the fact that points on boundaries between cells (epistemic states) A_i are (roughly) mapped onto points on boundaries between cells A_i. As a consequence, $\mathcal{P}_g$ is dynamically stable, the definition of the corresponding epistemic states is robust under the dynamics, and spurious correlations due to blurring cells are excluded.

The concept of a generating partition is related to the concept of a *Markov chain* in the theory of stochastic systems. Every deterministic system of first order gives rise to a Markov chain which is generally neither ergodic nor irreducible. Such Markov chains can be obtained by so-called *Markov partitions* that exist for expanding or hyperbolic dynamical systems (Sinai 1968, Bowen 1970, Ruelle 1989). For non-hyperbolic systems no corresponding existence theorem is available, and the construction can be even more tedious than for hyperbolic systems (Viana *et al.* 2003). For instance, both Markov and generating partitions for nonlinear systems are generally non-homogeneous, i.e. their cells are typically of different size and form.[2]

6   Symbolic Dynamics

Since generating partitions are stable under the phase space dynamics Φ, they can be used to construct symbol sequences s in a symbolic representation space

---

[2] Every Markov partition is generating, but the converse is not necessarily true (Crutchfield 1983, Crutchfield and Packard 1983). For the construction of generating partitions from empirical data it is often more convenient to approximate them by Markov partitions (Froyland 2001, Allefeld et al. 2009).

S in such a way that s is *topologically equivalent* to Φ.[3] This idea is exploited in the field of symbolic dynamics (Lind and Marcus 1995), where a continuous mapping $\pi : \Omega \to S$, called an *intertwiner*, is defined whose inverse π^{-1} exists and is also continuous. Then, the dynamics of epistemic states in Ω can be faithfully expressed as a symbol sequence $s \in S$ by:

$$\Phi = \pi \circ s \circ \pi^{-1} \qquad (10)$$

If the epistemic states A_i in Ω are cells of a generating partition, the intertwiner π exists, and s and Φ are guaranteed to be topologically equivalent. This means essentially that "neighboring" epistemic states in Ω will be mapped onto "neighboring" states in S. The construction of $\mathcal{P}_g$ entails that differences between epistemically equivalent ontic states in Ω are deliberately disregarded.

Partitions that are not generating lead to symbolic dynamics deviating from perfect topological equivalence. Skufca and Bollt (2008) investigated how the corresponding deviation of the map from Ω to S from an intertwiner can be characterized quantitatively by a "homeomorphic defect". This paves the way to specify the degree to which a symbolic description is a faithful representation of an underlying phase space dynamics.

Note that the concept of topological equivalence differs from *topological conjugacy* if the dynamics is continuous in time. Topological conjugacy requires an intertwiner mapping individual trajectories, i.e. ontic states defined pointwise in Ω, which can be parametrized pointwise in time. By contrast, epistemic states $A_i \in \Omega$ have no individual trajectories but sets of trajectories, so that π cannot map phase space states A_i onto symbolic states s together with a one-to-one mapping of their time parameter. This motivates topological equivalence as a relation weaker than topological conjugacy.

7   Improper Partitions

For improper partitions that are not generating, Bollt et al. (2001) coined the notion of "misplaced" partitions. Their use to determine the Kolmogorov-Sinai entropy leads to a systematic underestimation, because the cells of misplaced partitions are not stable under the dynamics and, thus, entail blurring effects of cell boundaries effectively violating the disjointness of epistemic states. As a consequence, there will be "spurious" correlations in addition to those originating from the dynamics itself. These "spurious" correlations obviously arise from epistemic states, not from decompositions of ontic entangled states. Therefore they differ drastically from entanglement correlations as exhibited by entangled quantum systems (cf. Atmanspacher and Primas 2003).

---

[3] While the construction of symbolic descriptions based on generating partitions is essentially motivated by their *stability* under the dynamics, a viable alternative relies on *information* theoretical ideas. This alternative is embedded in the framework of computational mechanics, as pioneered by Crutchfield and coworkers. See Crutchfield and Shalizi (2001) for a comprehensive review, and Shalizi and Moore (2003) for relations between their and our approach.

Although misplaced partitions are undesirable for extracting the Kolmogorov-Sinai entropy of a system or for defining faithful symbolic representations of the system dynamics by topologically equivalent symbol strings, they may be interesting for other purposes. For instance, they imply non-Boolean features arising from coarse grainings of purely classical phase spaces (cf. Westmoreland and Schumacher 1993). In other words, improper partitions may lead to a multitude of symbolic descriptions that are (all or partly) incompatible with each other, yet being (all or partly) necessary for a complete picture of the system considered.[4]

This may be a reason (surely not the only one) why sciences dealing with situations far more complex than in physics show a profound tendency toward non-universal theoretical frameworks of thinking. If phase space partitions of complex systems are set up ad hoc, the likelihood to find a proper (generating) partition is extremely low, and incompatible descriptions are an almost certain consequence. Atmanspacher and beim Graben (2007) argued along those lines for symbolic psychological descriptions and proposed a way to construct such descriptions based on proper partitions of neural phase spaces. A pertinent example of such a construction was demonstrated by Allefeld et al. (2009).

8   Compatibility and Other Relations Between Partitions

For a brief summary of possible relations between partitions we consider two observables f and g inducing epistemic states according to partitions $\mathcal{F}$ and $\mathcal{G}$. Then we can define the following relations (cf. beim Graben and Atmanspacher 2009).

- Two partitions $\mathcal{F}$ and $\mathcal{G}$ are *compatible* if and only if they are both generating, $R\mathcal{F} = R\mathcal{G} = \mathcal{I}$. This means that every ontic state x_o is epistemically accessible as a pre-image $\Phi^{-t}(\mathcal{F}, \mathcal{G})$.
- Two partitions $\mathcal{F}$ and $\mathcal{G}$ are *incompatible* if at least one of them is not generating, $R\mathcal{F} \neq R\mathcal{G}$.
- Two partitions $\mathcal{F}$ and $\mathcal{G}$ are *complementary, or maximally incompatible*, if their finest refinements are disjoint, $R\mathcal{F} \cap R\mathcal{G} = \emptyset$.
- Two partitions $\mathcal{F}$ and $\mathcal{G}$ are *comparable* if $R\mathcal{F}$ is a refinement of $R\mathcal{G}$ or vice versa. This entails that compatibility implies comparability. Even incompatible partitions may be comparable, if one of them is generating.
- Two partitions $\mathcal{F}$ and $\mathcal{G}$ are *commensurable* if a common language $T(\mathcal{U})$ embedding $T(\mathcal{F})$ and $T(\mathcal{G})$ exists (cf. Primas 1977) such that $R\mathcal{U}$ is a refinement of $R\mathcal{F}$ and $R\mathcal{G}$. Comparability implies commensurability.

9   Epistemic Entanglement

An interesting implication of improper, misplaced partitions is that they produce coarse grainings that change dynamically, thus yielding correlations in the dynamics of the system that are not a result of the dynamics itself but of overlapping

---

[4] Primas (2007) proposed the formal framework of *partial Boolean algebras* to refer to locally Boolean propositional lattices pasted together in a non-Boolean fashion.

coarse grains. For reasons mentioned above, such correlations are undesirable in symbolic dynamics and ergodic theory. However, they produce features that may look phenomenologically like entanglement correlations insofar as they are not explainable in terms of causal interactions of a system (e.g., with its environment).

This could provide insight concerning particular quantum-like features in classical systems, e.g. "Brownian entanglement" as reported by Allahverdian et al. (2005). Two particles undergoing Brownian motion were shown to create correlations analogous to quantum entanglement for coarse-grained velocities. From the perspective of our approach, it may be conjectured that this coarse-graining yields improper partitions inducing the correlations in question. Allahverdian et al.'s observation that the correlations disappear for an increasingly refined resolution of the coarse-graining points to an asymptotic epistemic accessibility of classical ontic particle states in their study.

Since ontic entanglement, as in genuinely entangled quantum systems, does not depend on measurement resolution or other partitioning issues, varying correlations due to altered partitions are a clear indicator for *epistemic entanglement*. This raises the question of whether it might be possible to adjust the degree of such epistemic entanglement in a controlled way. To our knowledge, this has not been studied so far, and at present we can only speculate about this possibility and its potential value. In the remaining sections we will sketch some corresponding ideas.

10   Acategorial Mental States

A state exhibiting epistemic entanglement according to blurred boundaries as discussed above would be a state represented by the intersection $A_i \cap A_j$ of non-disjoint states A_i and A_j. In a way, such a "superposition" state shares features of both A_i and A_j. On the other hand, neither A_i nor A_j is actualized because the actual state resides somehow "in between" them, offering the potential to actualize either one or the other state. Needless to say, this resembles the idea of a "reduction" of a quantum superposition state very closely.

An application of this idea to mental states was proposed by Atmanspacher (1992) and recently elaborated by Atmanspacher and Fach (2005) and Feil and Atmanspacher (2010). The present mainstream understanding of mental activity is framed by mental representations (or categories), which have been learned and stored, and which can be actualized by suitable stimuli (cf. Metzinger 2003). Mental states that actualize such representational categories are temporarily stable categorial states.

The notion of *acategorial states*, taken from Gebser (1986), has been used to address intermediate phases, for instance phases during which the mental state transits from one categorial state to another. The possibility of acategorial states depends crucially on the presence of established representations, none of which is actualized by an acategorial state though. While categorial states reside in stable mental representations strictly distinguishable from each other, inherently unstable acategorial states reside between adjacent categorial states and hold the possibility to relax into each one of them.[5]

Categorial states can be represented as epistemic states in appropriate phase spaces (Atmanspacher 1992, Feil and Atmanspacher 2010), and it is a challenging speculation to conceive of acategorial states as states exhibiting epistemic entanglement as indicated above. How might the experience of such states be like? A pertinent remark by Sudarshan (1983), responding to the question of how quantum states might be "perceived directly", proposes a mode of awareness in which

> "sensations, feelings, and insights are not neatly categorized into chains of thoughts, nor is there a step-by-step development of a logical-legal argument-to-conclusion. Instead, patterns appear, interweave, coexist; and sequencing is made inoperative. Conclusions, premises, feelings, and insights coexist in a manner defying temporal order."

11   Temporal Nonlocality

From a slightly different perspective, recent work by Atmanspacher and Filk (2010) on bistable perception suggests that the phenomenology described by Sudarshan (1983) may be related to the violation of temporal Bell inequalities entailing temporally nonlocal correlations.[6] It is a necessary condition for such a violation that the dynamics of the system considered is governed by operators that do not commute.

The resulting *temporal nonlocality* of mental states can be interpreted such that these states cannot be sharply (pointwise) localized along the time axis, and their characterization by sharp (classical) observable variables is inappropriate. Rather, temporally nonlocal states appear to be "stretched" over an extended time interval whose length may depend on the specific system considered. Within this interval, relations such as "earlier" or "later" are illegitimate designators of the system state. This is just another way of saying that it is impossible to define causal relationships within such a time interval (Filk and von Müller 2009).

It is tempting to relate this *temporal nonlocality* to a "window of temporal nowness", a concept that transcends a sharp boundary of presence between past and future (Filk and von Müller 2009, Pöppel 1997). However, the idea itself is much older and dates back at least to James' notion of the "specious present", a present mental state extending over a time interval rather than fixed to an instant of vanishing duration.

---

[5] By contrast, non-categorial states would be states without established representations. Feil and Atmanspacher (2010) suggested that acateogrial and non-categorial states are two different variants of the currently much discussed philosophical notion of "non-conceptual mental content" (Bermúdez and Cahen 2008).

[6] See also Atmanspacher and Filk (2011). While the original Bell inequalitites and their associated effects of nonlocality are usually discussed in terms of spatial relations between spatial subsystems, temporal Bell inequalities refer to relations between temporal segments of the history of a system.

Acategorial states are interesting candidates for temporal nonlocality as a property of mental states. Their intrinsic instability can easily be related to an indeterminate location in time that effectively amounts to their temporal extension. Presently we do not know whether and how it might be possible to actively control the temporal extent of such states. Considering them as epistemically entangled states according to Section 9 could provide theoretical access to this question.

References

1. Aerts, D., Durt, T., Grib, A., Van Bogaert, B., Zapatrin, A.: Quantum Structures in Macroscopical Reality. International Journal of Theoretical Physics 32, 489–498 (1993)
2. Allahverdyan, A.E., Khrennikov, A., Nieuwenhuizen, T.M.: Brownian entanglement. Physical Review A 72, 32102 (2005)
3. Allefeld, C., Atmanspacher, H., Wackermann, J.: Mental States as Macrostates Emerging from Brain Electrical Dynamics. Chaos 19, 015102 (2009)
4. Atmanspacher, H.: Categoreal and Acategoreal Representation of Knowledge. Cognitive Systems 3, 259–288 (1992)
5. Atmanspacher, H.: Quantum Approaches to Consciousness. In: Zalta, E. (ed.) Stanford Encyclopedia of Philosophy (2011)
6. Atmanspacher, H., beim Graben, P.: Contextual Emergence of Mental States from Neurodynamics. Chaos and Complexity Letters 2, 151–168 (2007)
7. Atmanspacher, H., Fach, W.: Acategoriality as Mental Instability. Journal of Mind and Behavior 26, 161–186 (2005)
8. Atmanspacher, H., Filk, T.: A Proposed Test of Temporal Nonlocality in Bistable Perception. Journal of Mathematical Psychology 54, 314–321 (2010)
9. Atmanspacher, H., Filk, T.: Contra Classical Causality: Violating Temporal Bell Inequalities in Mental Systems (to be published, 2011)
10. Atmanspacher, H., Primas, H.: Epistemic and Ontic Quantum Realities. In: Castell, L., Ischebeck, O. (eds.) Time, Quantum, and Information, pp. 301–321. Springer, Heidelberg (2003)
11. Atmanspacher, H., Römer, H., Walach, H.: Weak Quantum Theory: Complementarity and Entanglement in Physics and Beyond. Foundations of Physics 32, 379–406 (2002)
12. beim Graben, P., Atmanspacher, H.: Complementarity in Classical Dynamical Systems. Foundations of Physics 36, 291–306 (2006)
13. beim Graben, P., Atmanspacher, H.: Extending the Philosophical Signifance of the Idea of Complementarity. In: Atmanspacher, H., Primas, H. (eds.) Recasting Reality. Wolfgang Pauli's Philosophical Ideas and Contemporary Science, pp. 99–113. Springer, Berlin (2009)
14. Bermúdez, J.L., Cahen, A.: Nonconceptual mental content. Stanford Encyclopedia of Philosophy (2008),
    http://plato.stanford.edu/entries/content-nonconceptual/
15. Bollt, E., Stanford, T., Lai, Y., Życzkowski, K.: What Symbol Dynamics Do We Get with a Misplaced Partition? On the Validity of Threshold Crossings Analysis of Chaotic Time-Series. Physica D 154, 259–286 (2001)
16. Bowen, R.: Markov Partitions for Axiom A Diffeomorphisms. American Journal of Mathematics 92, 725–747 (1970)
17. Crutchfield, J.P.: Noisy Chaos. PhD thesis at the University of California at Santa Cruz, Sec. 5 (1983)

18. Crutchfield, J.P., Packard, N.H.: Symbolic Dynamics of Noisy Chaos. Physica D 7, 201–223 (1983)
19. Feil, D., Atmanspacher, H.: Acategorial States in a Representational Theory of Mental Processes. Journal of Consciousness Studies 17(5-6), 72–101 (2010)
20. Filk, T., von Müller, A.: Quantum Physics and Consciousness: The Quest for a Common Conceptual Foundation. Mind and Matter 7, 59–79 (2009)
21. Froyland, G.: Extracting Dynamical Behavior Via Markov Models. In: Mees, A.I. (ed.) Nonlinear Dynamics and Statistics, pp. 281–312. Birkhäuser, Boston (2001)
22. Gebser, J.: The Ever-Present Origin. Ohio University Press, Columbus (1986)
23. Harrigan, N., Spekkens, R.W.: Einstein, Incompleteness, and the Epistemic View of Quantum States. Foundations of Physics 40, 125–157 (2010)
24. Khrennikov, A.Y.: Classical and Quantum Mechanics on Information Spaces with Applications to Cognitive, Psychological, Social, and Anomalous Phenomena. Foundations of Physics 29, 1065–1098 (1999)
25. Kolmogorov, A.N.: New Metric Invariant of Transitive Dynamical Systems and Endomorphisms of Lebesgue Spaces. Doklady of Russian Academy of Sciences 124, 754–755 (1958)
26. Lind, D., Marcus, B.: An Introduction to Symbolic Dynamics and Coding. Cambridge University Press, Cambridge (1995)
27. Metzinger, T.: Being No One. MIT Press, Cambridge (2003)
28. Pöppel, E.: A Hierarchical Model of Temporal Perception. Trends in Cognitive Science 1, 56–61 (1997)
29. Primas, H.: Mathematical and Philosophical Questions in the Theory of Open and Macroscopic Quantum Systems. In: Miller, A.I. (ed.) Sixty-Two Years of Uncertainty, pp. 233–257. Plenum, New York (1990)
30. Primas, H.: Theory Reduction and Non-Boolean Theories. Journal of Mathematical Biology 4, 281–301 (1977)
31. Primas, H.: Non-Boolean Descriptions for Mind-Matter Problems. Mind and Matter 5, 7–44 (2007)
32. Ruelle, D.: The Thermodynamic Formalism for Expanding Maps. Communications of Mathematical Physics 125, 239–262 (1989)
33. Shalizi, D.R., Crutchfield, J.P.: Computational Mechanics: Pattern and Prediction, Structure and Simplicity. Journal of Statistical Physics 104, 817–879 (2001)
34. Shalizi, C.R., Moore, C.: What Is a Macrostate? Subjective Observations and Objective Dynamics (2003) (preprint),
http://philsci-archive.pitt.edu/1119/1/whats-macro.pdf
35. Sinai, Y. G.: On the Notion of Entropy of a Dynamical System. Doklady of Russian Academy of Sciences 124, 768–771 (1959)
36. Sinai, Y. G.: Markov Partitions and C-Diffeomorphisms. Functional Analysis and its Applications 2, 61–82 (1968)
37. Skufca, J.D., Bollt, E.M.: A concept of homeomorphic defect for defining mostly conjugate dynamical systems. Chaos 18, 013118 (2008)
38. Spekkens, R.W.: In Defense of the Epistemic View of Quantum States: A Toy Theory. Physical Review 75, 032110 (2007)
39. Sudarshan, E.C.G.: Perception of Quantum Systems. In: van der Merwe, A. (ed.) Old and New Questions in Physics, Cosmology, Philosophy, and Theoretical Biology, pp. 457–467. Plenum, New York (1983)
40. Viana, R.L., Grebogi, G., de Pinto, S.E., Barbosa, J.R.R., Grebogi, C.: Pseudo-Deterministic Chaotic Systems. International Journal of Bifurcation and Chaos 13, 3235–3253 (2003)
41. Westmoreland, M.D., Schumacher, B.D.: Non-Boolean Derived Logics for Classical Systems. Physical Review A 48, 977–985 (1993)

Quantum Structure in Cognition: Why and How Concepts Are Entangled

Diederik Aerts and Sandro Sozzo

Center Leo Apostel for Interdisciplinary Studies
Vrije Universiteit Brussel
Krijgskundestraat 33, B-1160 Brussels, Belgium
{diraerts,ssozzo}@vub.ac.be

Abstract. One of us has recently elaborated a theory for modelling concepts that uses the state context property (SCoP) formalism, i.e. a generalization of the quantum formalism. This formalism incorporates context into the mathematical structure used to represent a concept, and thereby models how context influences the typicality of a single exemplar and the applicability of a single property of a concept, which provides a solution of the *Pet-Fish problem* and other difficulties occurring in concept theory. Then, a quantum model has been worked out which reproduces the membership weights of several exemplars of concepts and their combinations. We show in this paper that a further relevant effect appears in a natural way whenever two or more concepts combine, namely, *entanglement*. The presence of entanglement is explicitly revealed by considering a specific example with two concepts, constructing some Bell's inequalities for this example, testing them in a real experiment with test subjects, and finally proving that Bell's inequalities are violated in this case. We show that the intrinsic and unavoidable character of entanglement can be explained in terms of the weights of the exemplars of the combined concept with respect to the weights of the exemplars of the component concepts.

Keywords: Concept combination, Bell's inequalities, entanglement, quantum cognition.

1 Introduction

Understanding the mechanism of how concepts combine to form sentences and texts so that it is possible to communicate meaning among human minds is one of the major challenges in the psychological studies on human thought. None of the existing theories on concepts explains however 'how concepts combine'. This *combination problem* was manifestly revealed by Hampton's experiments [1,2] which measured the deviation from classical set theoretic membership weights of exemplars with respect to pairs of concepts and their conjunction or disjunction. Hampton's investigation was motivated by the so-called *Guppy effect* in concept conjunction found by Osherson and Smith [3]. These authors considered the concepts *Pet* and *Fish* and their conjunction *Pet-Fish*, and observed that, while an exemplar such as *Guppy* was a very typical example of *Pet-Fish*, it was

D. Song et al. (Eds.): QI 2011, LNCS 7052, pp. 116–127, 2011.

neither a very typical example of *Pet* nor of *Fish*. Hence, the typicality of a specific exemplar with respect to the conjunction of concepts shows an unexpected behavior from the point of view of classical set and probability theory. As a result of this work, the problem is often referred to as the *Pet-Fish problem* and the effect is usually called the *Guppy effect*. Hampton identified a Guppy-like effect for the membership weights of exemplars with respect to pairs of concepts and their conjunction [1], and equally so for the membership weights of exemplars with respect to pairs of concepts and their disjunction [2]. Several experiments have since been performed (see, e.g., [4]) and many approaches have been propounded to solve the Pet-Fish problem (see, e.g., fuzzy set based theories [5,6,7]) and to provide a satisfactory mathematical model of concept combinations (see, e.g., explanation based theories [8,9,10]). But none of the currently existing concept theories provide a satisfactory description or explanation of such effects [4,9,10].

Inspired by a formalism providing an operational foundation of quantum mechanics [11,12,13,14], one of the authors has elaborated, together with some co-workers, a *State Context Property* (*SCoP*) formalism to model and represent concepts [15,16,17,18]. In the SCoP formalism each concept is associated with well defined sets of states, contexts and properties. Concepts change continuously under the influence of a context and this change is described by a change of the state of the concept. For each exemplar of a concept, the typicality varies with respect to the context that influences it. Analogously, for each property, the applicability varies with respect to the context. This implies the presence of both a *contextual typicality* and an *applicability effect*. The Pet-Fish problem is solved in the SCoP formalism because in different combinations the concepts are in different states. In particular, in the combination *Pet-Fish* the concept *Pet* is in a state under the context *The Pet is a Fish*. The state of *Pet* under the context *The Pet is a Fish* has different typicalities, which explains the Guppy effect. On the basis of the SCoP formalism, a mathematical model using the formalism of quantum mechanics in Hilbert space has been worked out which allows one to reproduce the experimental results obtained by Hampton on conjunctions and disjunctions of concepts. This formulation identifies the presence of typically quantum effects in the mechanism of combination of concepts, e.g., contextual influence, superposition, interference, emergence, etc. [19,20,21,22,23,24,25].

In this paper we show that another relevant effect which is usually considered as characteristic of quantum mechanical entities, that is, *entanglement*, is present whenever two or more concepts combine. The presence of entanglement is explicitly revealed by considering two concepts, i.e. *Animal* and *Acts*, and their combination *The Animal Acts*, together with some exemplars *Horse, Bear, Tiger, Cat* (for *Animal*) and *Growls, Whinnies, Snorts, Meows* (for *Acts*), and constructing some Bell's inequalities in the version derived by Clauser, Horne, Shimony and Holt [26] (Sec. 2). We then test these Bell's inequalities in a real experiment with 81 test subjects and analyze the obtained data (Sec. 3). The experiment shows a significant violation of Bell's inequalities, hence it proves the entanglement between the concepts *Animal* and *Acts* when they form the sentence *The Animal Acts* (by the term *entanglement* we actually mean the presence of nonclassical correlations violating Bell's inequalities, without reference to any mathematical representation in Hilbert spaces). Moreover, we compare the obtained data with the results that would have been obtained if context and meaning had not influenced the subjects' minds. In the latter case, indeed, Bell's inequalities are not violated, hence their violation in our experiment shows that meaning and context play a basic role in the combination of concepts. We finally provide an explanation of the origins and ubiquity of entanglement in combined concepts in terms of weights of the exemplars of the combined concept with respect to the weights of the exemplars of the component concepts (Sec. 4).

We conclude this section by observing that the potentially fundamental role played by entanglement in word association was pointed out by Nelson and McEvoy and Bruza et al. in [27,28]. In [29] it is shown that if one assumes that words can become entangled in the human mental lexicon, then one can provide a unified framework in which two seemingly competing approaches for modeling the activation level of words in human memory, namely, the *Spreading Activation* and the *Spooky-activation-at-a-distance*, can be recovered.

2 Detecting Entanglement between Concepts

We illustrate in this section how entanglement appears in a natural way whenever two or more concepts combine. To this aim, we analyze here an example with two concepts and a combination along the lines put forward in [16,17,18].

We regard the sentence *The Animal Acts* as a conceptual entity, hence as a combination of the concepts *Animal* and *Acts*. Then, we show the presence of entanglement between these two concepts by testing Bell's inequality with respect to them. We consider two couples of exemplars or states of the concept *Animal*, namely *Horse*, *Bear* and *Tiger*, *Cat*, and also two couples of exemplars or states of the concept *Acts*, namely *Growls*, *Whinnies* and *Snorts*, *Meows* – for our experiment we specifically consider forms of actions, hence exemplars of *Acts*, which consists of possible animal sounds, hence exemplars of *Making A Sound*. Our first experiment A consists in test subjects choosing between the two exemplars *Horse* and *Bear* to answer the question 'is a good example of' the concept *Animal*, and we put $E(A) = +1$ if *Horse* is chosen, hence the state of *Animal* changes to *Horse*, and $E(A) = -1$ if *Bear* is chosen, hence the state of *Animal* changes to *Bear*, introducing in this way the function E which measures the 'expectation value' for the test outcomes concerned. Our second experiment A' consists in test subjects choosing between the two exemplars *Tiger* and *Cat* to answer the question 'is a good example of' the concept *Animal*, and we consistently put $E(A') = +1$ if *Tiger* is chosen and $E(A') = -1$ if *Cat* is chosen to introduce a measure of the expectation value. The third experiment B consists in test subjects choosing between the two exemplars *Growls* and *Whinnies* to answer the question 'is a good example of' the concept *Acts*, with $E(B) = +1$ if *Growls* is chosen and $E(B) = -1$ if *Whinnies* is chosen. The fourth experiment B' consists in test subjects choosing between the exemplars *Snorts* and *Meows* to answer the question 'is a good example of' the concept *Acts*, with $E(B') = +1$ if *Snorts* is chosen and $E(B') = -1$ if *Meows* is chosen.

Let us now consider coincidence experiments in combinations AB, $A'B$, AB' and $A'B'$ for the conceptual combination *The Animal Acts*. Concretely, this means that, for example, test subjects taking part in the experiment AB, to answer the question 'is a good example of', will choose between the four possibilities (1) *The Horse Growls*, (2) *The Bear Whinnies* – and if one of these is chosen we put $E(AB) = +1$ – and (3) *The Horse Whinnies*, (4) *The Bear Growls* – and if one of these is chosen we put $E(AB) = -1$. For the coincidence experiment, $A'B$ subjects, to answer the question 'is a good example of', will choose between (1) *The Tiger Growls*, (2) *The Cat Whinnies* – and in case one of these is chosen we put $E(A'B) = +1$ – and (3) *The Tiger Whinnies*, (4) *The Cat Growls* – and in case one of these is chosen we put $E(A'B) = -1$. For the coincidence experiment, AB' subjects, to answer the question 'is a good example of', choose between (1) *The Horse Snorts*, (2) *The Bear Meows* – and in case one of these is chosen we put $E(AB') = +1$ – and (3) *The Horse Meows*, (4) *The Bear Snorts* – and in case one of these is chosen we put $E(AB') = -1$. And finally, for the coincidence experiment, $A'B'$ subjects, to answer the question 'is a good example of', will choose between (1) *The Tiger Snorts*, (2) *The Cat Meows* – and in case one of these is chosen we put $E(A'B') = +1$ – and (3) *The Tiger Meows*, (4) *The Cat Snorts* – and in case one of these is chosen we put $E(A'B') = -1$. We can now evaluate the expectation values $E(A', B')$, $E(A', B)$, $E(A, B')$ and $E(A, B)$ associated with the coincidence experiments $A'B'$, $A'B$, AB' and AB, respectively, and substitute them into the Clauser-Horne-Shimony-Holt variant of Bell's inequality [26]

$$-2 \leq E(A', B') + E(A', B) + E(A, B') - E(A, B) \leq 2. \tag{1}$$

From the well-known analysis of Bell's inequality follows that in case the experimental expectation values violate Eq. (1), a local and classical probabilistic description is not possible, and entanglement exists between the given concepts. Thus, by the sentence Animal *is entangled with* Acts we mean the experimental fact that these two concepts exhibit nonclassical correlations, without referring to any mathematical representation in Hilbert spaces. The connections with entangled states in tensor product Hilbert spaces will be outlined in Sec. 4.

We note that the maximum violation of the Bell's inequality in Eq. (1) occurs when the quantity $E(A', B') + E(A', B) + E(A, B') - E(A, B)$ is equal to $+4$, that is, when the outcome for each one of the members of this expression is $+1$, $+1$, $+1$ and -1, respectively. Let us make an intuitive analysis of the situation such that we can see why Bell's inequality will most probably be violated for our experiment. In the coincidence experiment AB, both *The Horse Whinnies* and *The Bear Growls* will yield rather high scores, with the two remaining possibilities *The Horse Growls* and *The Bear Whinnies* being chosen little. This means that we will get $E(A, B)$ close to -1. On the other hand, in the coincidence experiment $A'B$ one of the four choices will be prominent, namely *The Tiger Growls*, while the three other possibilities, *The Cat Whinnies*, *The Tiger Whinnies*, and *The Cat Growls*, will be much less present amongst the choices made by the test subjects. This means that we have $E(A', B)$ close to $+1$. In the two remaining coincidence experiments, we equally have that only one of the choices is prominent. For A, B', this is *The Horse Snorts*, with the other three *The Bear Meows*, *The Horse Meows* and *The Bear Snorts* being much less present. For $A'B'$, the prominent choice is *The Cat Meows*, while the other three *The Tiger Snorts*, *The Tiger Meows* and *The Cat Snorts* are much less present. This means that we have $E(A, B')$ is close to $+1$ and $E(A', B')$ is close to $+1$. Coming to the expectation values, we hence can expect that Eq. (1) be violated, and that case (ii) occurs such that the existence of entanglement between the considered concepts would be proven.

One of us has recently shown [30] that Eq. (1) is violated in the concept combination *The Animal Acts* by using the World Wide Web as a conceptual domain. In the next section we will show that a violation occurs also when the data are collected from a real experiment with test subjects following the standard procedure of psychology experiments in concept research.

3   Description of the Experiment

The entanglement mentioned in the foregoing section was tested in an experiment where 81 participating subjects were presented with a questionnaire to be filled out accompanied by the following text:

This study has to do with what we have in mind when we use words that refer to categories, and more specifically 'how we think about examples of categories'. Let us illustrate what we mean. Consider the category 'fruit'. Then 'orange' and 'strawberry' are two examples of this category, but also 'fig' and 'olive' are examples of the same category. In each test of the questionnaire you will be asked to pick one of the examples of a set of given examples for a specific category. And we would like you to pick that example that you find 'a good example' of the category. In case there are more than one example which you find a good example, pick then the one you find the best of all the good examples. In case there are two examples which you both find equally good, and hence hesitate which ones to take, just take then the one you slightly prefer, however slight the preference might be. It is mandatory that you always 'pick one and only one example', hence in case of doubt, anyhow pick one and only one example. This is necessary for the experiment to succeed. So, one of the tests could be that the category 'fruit' is given, and you are asked to pick one of the examples 'orange', 'strawberry', 'fig' or 'olive' as a good example, and in case of doubt the best of the ones you doubt about, and in case you cannot decide, pick one anyhow. Let all aspects of yourself play a role in the choice you make, ratio, but also imagination, feeling, emotion, and whatever.

Let us now examine the obtained results.

For the coincidence experiment AB, 4 subjects chose the example *The Horse Growls* as a good example of the combination *The Animal Acts*, 5 subjects chose *The Bear Whinnies*, 51 subjects chose *The Horse Whinnies*, and 21 subjects chose *The Bear Growls*. This means that on a totality of 81 test subjects we get fractions of 4, 5, 51 and 21 for the different combinations considered. This allows us to calculate the probability for one of the combinations to be chosen. We have $P(A_1, B_1) = 4/81 = 0.0494$ for *The Horse Growls*, $P(A_2, B_2) = 21/81 = 0.2593$ for *The Bear Whinnies*, $P(A_1, B_2) = 51/81 = 0.6296$ for *The Horse Whinnies* and $P(A_2, B_1) = 5/81 = 0.0617$ for *The Bear Growls*. Knowing these probabilities, we can again calculate the expectation value for this coincidence experiment by means of the equation $E(A, B) = P(A_1, B_1) + P(A_2, B_2) - P(A_2, B_1) - P(A_1, B_2) = -0.7778$. We calculate the expectation values $E(A', B)$, $E(A, B')$ and $E(A', B')$ in an analogous way. For the coincidence experiment $A'B$, 63 subjects chose the example *The Tiger Growls* as a good example of the combination *The Animal Acts*, 4 subjects chose *The Cat Whinnies*, 7 subjects chose *The Tiger Whinnies*, and 7 subjects chose *The Cat Growls*. This gives $P(A_1', B_1) = 0.7778$, $P(A_2', B_2) = 0.0494$, $P(A_1', B_2) = 0.0864$ and $P(A_2', B_1) = 0.0864$, hence $E(A', B) = 0.6543$. For the coincidence experiment AB', 48 subjects chose the example *The Horse Snorts* as a good example of the combination *The Animal Acts*, 7 subjects chose *The Bear Meows*, 2 subjects chose *The Horse Meows*, and 24 subjects chose *The Bear Snorts*. This gives $P(A_1, B_1') = 0.5926$, $P(A_2, B_2') = 0.0864$, $P(A_1, B_2') = 0.0247$ and $P(A_2, B_1') = 0.2963$, hence $E(A, B') = 0.3580$. For the coincidence experiment $A'B'$, 12 subjects chose the example *The Tiger Snorts* as a good example of the combination *The Animal Acts*, 54 subjects chose *The Cat Meows*, 7 subjects chose *The Tiger Meows*, and 8 subjects chose *The Cat Snorts*. This gives $P(A_1', B_1') = 0.1481$, $P(A_2', B_2') = 0.6667$, $P(A_1', B_2') = 0.0864$ and $P(A_2', B_1') = 0.0988$, hence $E(A', B') = 0.6296$. For the expression appearing in the Clauser-Horne-Shimony-Holt variant of Bell's inequalities, we get

$$E(A', B') + E(A', B) + E(A, B') - E(A, B) = 2.4197 \qquad (2)$$

which is manifestly greater than 2, hence it violates Bell's inequalities and reveals entanglement between the concept *Animal* and the concept *Acts*.

The above violation of Bell's inequalities constitutes our main result in this paper and we will exhaustively comment on it in the next section. But we first want to consider Bell's inequalities under different perspectives.

Suppose that there are two separated sources of knowledge, e.g., two test subjects, and consider the coincidence experiment AB. Let $P(A_1)$ be the probability that the first subject choose the exemplar *Horse* as a good example of the concept *Animal*, let $P(B_1)$ be the probability that the second subject choose the exemplar *Growls* as a good example of the concept *Acts*, and let us estimate the probability that the example *The Horse Growls* be a good example of the conceptual combination *The Animal Acts* as the product $P(A_1)P(B_1)$, that is, as the joint probability $P_{prod}(A_1, B_1)$ that the first subject choose *Horse* and the second subject choose *Growls*. By referring to the experimental data that have been collected we have $P(A_1) = 43/81 = 0.5309$, $P(B_1) = 39/81 = 0.4815$, $P_{prod}(A_1, B_1) = P(A_1)P(B_1) = 0.2556$. Analogously, we can calculate the probability that *The Bear Whinnies* be a good example of *The Animal Acts* as the product of the probability $P(A_2)$ that the first subject choose *Bear* as a good example of *Animal* times the probability $P(B_2)$ that the second subject choose *Whinnies* as a good example of *Acts*. We find from empirical data $P(A_2) = 38/81 = 0.4691$,

$P(B_2) = 42/81 = 0.5185$, hence $P_{prod}(A_2, B_2) = P(A_2)P(B_2) = 0.2433$. By proceeding in an analogous way we calculate the probability that *The Horse Whinnies* be a good example of *The Animal Acts* as the product of the probability $P(A_1)$ that the first subject choose *Horse* as a good example of *Animal* times the probability $P(B_2)$ that the second subject choose *Whinnies* as a good example of *Acts*. We find $P_{prod}(A_1, B_2) = P(A_1)P(B_2) = 0.2753$. Furthermore, if we calculate the probability that *The Bear Growls* be a good example of *The Animal Acts* as the product of the probability $P(A_2)$ that the first subject choose *Bear* as a good example of *Animal* times the probability $P(B_1)$ that the second subject choose *Growls* as a good example of *Acts*, we find $P_{prod}(A_2, B_1) = P(A_2)P(B_1) = 0.2259$. The expectation value is $E_{prod}(A, B) = P_{prod}(A_1, B_1) + P_{prod}(A_2, B_2) - P_{prod}(A_2, B_1) - P_{prod}(A_1, B_2) = -0.0022$. Let us now consider the coincidence experiment $A'B$. The probability that the first subject choose the example *Tiger* as a good example of *Animal* is $P(A_1') = 59/81 = 0.7284$, while the probability that the first subject choose *Cat* as a good example of *Animal* is $P(A_2') = 22/81 = 0.2716$. If we calculate the probability that *The Tiger Growls* be a good example of *The Animal Acts* as the product of the probability $P(A_1')$ that the first subject choose *Tiger* as a good example of *Animal* times the probability $P(B_1)$ that the second subject choose *Growls* as a good example of *Acts*, we find $P_{prod}(A_1', B_1) = P(A_1')P(B_1) = 0.3507$. Analogously, we find $P_{prod}(A_2', B_2) = P(A_2')P(B_2) = 0.1408$, $P_{prod}(A_1', B_2) = P(A_1')P(B_2) = 0.3777$, $P_{prod}(A_2', B_1) = P(A_2')P(B_1) = 0.1308$. The expectation value is $E_{prod}(A', B) = P_{prod}(A_1', B_1) + P_{prod}(A_2', B_2) - P_{prod}(A_2', B_1) - P_{prod}(A_1', B_2) = -0.0169$. Let us come to the coincidence experiment AB'. The probability that the second subject choose the example *Snorts* as a good example of *Acts* is $P(B_1') = 26/81 = 0.3210$, while the probability that the second subject choose *Meows* as a good example of *Acts* is $P(B_2') = 55/81 = 0.6790$. If we calculate the probability that *The Horse Snorts* be a good example of *The Animal Acts* as the product of the probability $P(A_1)$ that the first subject choose *Horse* as a good example of *Animal* times the probability $P(B_1')$ that the second subject choose *Snorts* as a good example of *Acts*, we find $P_{prod}(A_1, B_1') = P(A_1)P(B_1') = 0.1704$. Analogously, we find $P_{prod}(A_1, B_1') = P(A_2)P(B_2') = 0.3185$, $P_{prod}(A_1, B_1') = P(A_1)P(B_2') = 0.3605$, $P_{prod}(A_1, B_1') = P(A_2)P(B_1') = 0.1506$. The expectation value is $E_{prod}(A, B') = P_{prod}(A_1, B_1') + P_{prod}(A_2, B_2') - P_{prod}(A_2, B_1') - P_{prod}(A_1, B_2') = -0.0221$. Finally, let us consider the coincidence experiment $A'B'$. If we calculate the probability that *The Tiger Snorts* be a good example of *The Animal Acts* as the product of the probability $P(A_1')$ that the first subject choose *Tiger* as a good example of *Animal* times the probability $P(B_1')$ that the second subject choose *Snorts* as a good example of *Acts*, we find $P_{prod}(A_1', B_1') = P(A_1')P(B_1') = 0.2338$. Analogously, we find $P_{prod}(A_2', B_2') = P(A_2')P(B_2') = 0.1844$, $P_{prod}(A_1', B_2') = P(A_1')P(B_2') = 0.4946$, $P_{prod}(A_2', B_1') = P(A_2')P(B_1') = 0.0871$. The expectation value is $E_{prod}(A', B') = P_{prod}(A_1', B_1') + P_{prod}(A_2', B_2') - P_{prod}(A_2', B_1') - P_{prod}(A_1', B_2') = -0.1635$. The 'product' expectation values $E_{prod}(A, B)$, $E_{prod}(A', B)$, $E_{prod}(A, B')$ and $E_{prod}(A', B')$ can then be put into the Bell inequality, which gives

$$E_{prod}(A', B') + E_{prod}(A', B) + E_{prod}(A, B') - E_{prod}(A, B) = -0.2003. \quad (3)$$

This result is very different from the earlier obtained expression, and also does not violate Bell's inequalities. The reason for this is that in the case of 'separated sources of knowledge', the non-violation of Bell's inequalities is structural [30]. This statement can be proved as follows.

Lemma. If x, x', y and y' are real numbers such that $-1 \leq x, x', y, y \leq +1$ and $S = xy + xy' + x'y - x'y'$ then $-2 \leq S \leq +2$.

Proof. Since S is linear in all four variables x, x', y, y', it must take on its maximum and minimum values at the corners of the domain of this quadruple of variables, that is, where each of x, x', y, y' is +1 or -1. Hence at these corners S can only be an integer between -4 and +4. But S can be rewritten as $(x + x')(y + y') - 2x'y'$, and the two quantities in parentheses can only be 0, 2, or -2, while the last term can only be -2 or +2, so that S cannot equal -3, +3, -4, or +4 at the corners.

Since in the situation considered we have $P_{prod}(A_i, B_j) = P(A_i)P(B_j)$, $P_{prod}(A'_i, B_j) = P(A'_i)P(B_j)$, $P_{prod}(A_i, B'_j) = P(A_i)P(B'_j)$ and $P_{prod}(A'_i, B'_j) = P(A'_i)P(B'_j)$, we get $E(A, B) = E(A)E(B)$, $E(A', B) = E(A')E(B)$, $E(A, B') = E(A)E(B')$ and $E(A', B') = E(A')E(B')$, and hence from the lemma it follows that $-2 \leq E(A'B') + E(A'B) + E(AB') - E(AB) \leq +2$, which proves the Clauser-Horne-Shimony-Holt variant of Bell's inequalities to be valid.

The foregoing considerations show that one of the elements in the violation of Bell's inequalities is the non-product nature of the probabilities $P(A_i, B_j)$, $P(A'_i, B_j)$, $P(A_i, B'_j)$ and $P(A'_i, B'_j)$, e.g., $P(A_i, B_j) \neq P(A_i)P(B_j)$. If we understand why these coincidence probabilities are not of the product nature we can get an insight into one of the elements of the violation of Bell's inequalities for the situations that we have considered. Indeed, consider for example the probability $P(A_1, B_1)$ and let us analyze why it is different from $P(A_1)P(B_1)$. We have that $P(A_1, B_1)$ is the probability, empirically estimated, that a given test subject choose the sentence *The Horse Growls* as a good example of the concept *The Animal Acts*, and then we find $P(A_1, B_1) = 0.0494$. On the contrary, $P(A_1)P(B_1)$ is the probability that, of two given test subjects, the first choose *Horse* as a good example of *Animal* and the other choose independently *Growls* as a good example of *Acts*, and then we find $P(A_1)P(B_1) = 0.2556$. These values are very different. Indeed, the probability to find the sentence part *The Horse Growls* is little, for any meaning this sentence may have will not be easily ascertained, since it is most unusual for horses to growl. If however two 'separated' or 'independent' subjects are chosen at random, the probability that *Horse* be chosen by one of them, and *Growls* be chosen by the other, is substantial. The fundamental reason for this difference is that in the second case the choices are 'separated' or 'independent' or, rather, 'not connected by meaning'.

The results above show that 'meaning' plays a fundamental role in determining the experimental weights of the examples of concept combinations. But, there are stronger arguments to maintain that context and meaning are crucial in human thought, hence a combination of concepts is not like a 'bag of words', as implied by the mathematical structure of existing semantic theories, e.g., LSA.

Let us calculate the data that would have been obtained if the minds of the test subjects had not been influenced by context and meaning. Consider the coincidence experiment AB and suppose that a given subject chooses the exemplar *Horse* as a good example of the concept *Animal* and *Growls* as a good example of the concept *Acts*. Should context and meaning not play any role, then the subject would choose with certainty the example *The Horse Growls* as a good example of the combination *The Animal Acts*. We can thus evaluate the probability $P_{class}(A_1, B_1)$ that a given subject choose *Horse* in the experiment A and *Growls* in the experiment B. It is given by $P_{class}(A_1, B_1) = 19/81 = 0.2346$, where 19 is the number of subjects who chose *Horse* in the experiment A and *Growls* in the experiment B. This probability can be used as an estimation of the probability that a given subject choose *The Horse Growls* as a good example of *The Animal Acts*. We can repeat the same reasoning for the other possible results in the coincidence experiment AB, thus getting $P_{class}(A_2, B_2) = 0.2222$, $P_{class}(A_1, B_2) = 0.2963$ and $P_{class}(A_2, B_1) = 0.2469$. Hence the expectation value is $E_{class}(A, B) = P_{class}(A_1, B_1) + P_{class}(A_2, B_2) - P_{class}(A_1, B_2) - P_{class}(A_2, B_1) = -0.0864$ in this case. Analogously, we get $E_{class}(A', B) = 0.1235$, $E_{class}(A, B') = -0.0123$ and $E_{class}(A', B') = -0.1111$ for the expectation values of the other coincidence experiments. The 'classical' expectation values $E_{class}(A, B)$, $E_{class}(A', B)$, $E_{class}(A, B')$ and $E_{class}(A', B')$ can then be inserted into the Bell inequality, which gives

$$E_{class}(A', B') + E_{class}(A', B) + E_{class}(A, B') - E_{class}(A, B) = 0.0864. \quad (4)$$

As we can see, the obtained value does not violate Bell's inequalities. As a consequence, the violation of Bell's inequalities in the experiment that we have considered can be interpreted as proving that meaning and context are fundamental for the mechanism of construction of sentences.

To conclude this section we observe that we also performed a statistical analysis of the empirical data using the 't-test for paired two samples for means' to estimate the probability that the shifts from Bell's inequalities be due to chance. We compared the data collected in the real experiment with the data collected in the 'classical' experiment, where no influence of context and meaning is present. For the 16 pairs to compare the p-values came out as follows: 0.000392657, 0.003921785, 2.50665E-06, 0.820174295, 3.8846E-08, 0.011513803, 4.78134E-05, 0.741136115, 2.35428E-08, 0.000152291, 1.3612E-08, 0.006518053, 0.073431676, 7.38957E-12, 3.8846E-08, 0.56693215. This makes it possible to conclude convincingly that the deviation effects are not caused by random fluctuations.

4   Explanation of Entanglement in Concepts

A fundamental consequence of the experimental results obtained in Sec. 3 is that any formalism aiming at representing concepts should incorporate the possibility of having entangled concepts from the very beginning. In order to understand in depth the mechanism of entanglement between concepts together with the causes of its ubiquity, we put forward an analysis of the situation in this section with the aim of grasping the core element of entanglement for concept combinations.

Consider the concept *Animal*. This concept is an 'abstraction' of possible concrete exemplars of *Animal*, e.g., *Horse*, *Bear*, *Tiger*, *Cat*, etc. When we ask a subject to estimate whether a given example, say, *Horse* is a 'good example' of the concept *Animal* this operation corresponds to 'wandering into the realm of abstraction and concretization'. The concept *Animal* is then connected with the exemplars of *Animal* by weights, expressing frequencies of appearance and/or typicalities of the different exemplars. Analogously, the concept *Acts* is an abstraction of possible concrete exemplars of *Acts*, and also connected to these different exemplars by weights, expressing frequencies of appearance and/or typicalities. Let us now consider the concept *The Animal Acts* which is the combination of *Animal* and *Acts*. This is also an abstraction of possible exemplars. In the situation that we considered for the experiment the concrete exemplars are *The Horse Growls*, *The Tiger Meows*, etc. But, the weight of, say, *The Horse Growls* is not the product of the weight of *Horse* in *Animal* times the weight of *Growls* in *Acts* in this case, otherwise Bell's inequalities would have been satisfied. It follows that the essential element being at the origin of entanglement is that 'when concepts combine they do this inside the realm of where they exist as abstractions'. With other words, the combination *The Animal Acts*, is a combination of two abstractions *Animal* and *Acts*, but it does not connect with the concrete elements, i.e. the exemplars of *Animal* and *Acts*. No, it connects with its own set of exemplars, such a *The Horse Whinnies* or *The Bear Growls*, etc., which are in themselves combinations of exemplars of the original concepts, but even this is not necessary, also completely new exemplars can be considered for the combination. This is a very different way of combining than for example the way in which two classical physical object combine. Hence, entanglement is a direct and deep consequence of this special way of combining, for each combination choosing its own set of new exemplars, 'with new weight specifically linked to the individual exemplars', and not connecting to the product set of the old exemplars and corresponding weights. That concepts have this special way of combining in common with quantum entities might not be a coincidence, a hypothesis investigated in [19].

A consequence of the above analysis is that entanglement in concepts does not strictly depend on the linearity of the tensor product Hilbert space that can be used to model the entity *The Animal Acts* – we remind that the *Tsirelson inequalities* [31] hold in the specific case that we have considered, therefore a purely quantum model can be worked out in this case. Moreover, the type of model in Hilbert space that we would expect is the following. Let us denote the states of the concepts *Animal* and *Acts* by the unit vectors $|p_{Animal}\rangle$ and $|p_{Acts}\rangle$, respectively. Since *Animal* and *Acts* are both abstractions of, say, *Horse* and *Bear* and of *Growls* and *Whinnies*, respectively, we have

$$|p_{Animal}\rangle = a_1|p_H\rangle + a_2|p_B\rangle, \quad |p_{Acts}\rangle = b_1|p_G\rangle + b_2|p_W\rangle \tag{5}$$

where $|a_1|^2$ and $|a_2|^2$, and $|b_1|^2$ and $|b_2|^2$, respectively, are the weights that both concretizations carry, and the unit vectors $|p_H\rangle$, $|p_B\rangle$, $|p_G\rangle$ and $|p_W\rangle$ represent the states p_{Horse}, p_{Bear}, p_{Growls} and $p_{Whinnies}$, respectively. The ground state $p_{The\ Animal\ Acts}$ of the combination *The Animal Acts*, being an abstraction of 'all combinations of the concrete cases', is then represented by the unit vector

$$|p_{The\ Animal\ Acts}\rangle = c_1|p_{HG}\rangle + c_2|p_{BW}\rangle + c_3|p_{HW}\rangle + c_4|p_{BG}\rangle, \tag{6}$$

where the unit vectors $|p_{HG}\rangle$, $|p_{BW}\rangle$, $|p_{HW}\rangle$ and $|p_{BG}\rangle$ represent the states $p_{The\ Horse\ Growls}$, $p_{The\ Bear\ Whinnies}$, $p_{The\ Horse\ Whinnies}$ and $p_{The\ Bear\ Growls}$, respectively. Eq. (6) is not, in general, a product, hence it is not equal to the tensor product $|p_{Animal}\rangle \otimes |p_{Acts}\rangle$, which is the mathematical basis of the presence of entanglement.

The unavoidability of entanglement could explain the difficulties that scholars encounter in putting forward a modeling scheme for concepts and their combinations in which individual concepts are represented by a unique mathematical structure, e.g., vectors such as in LSA, without introducing the tensor product structure (see, e.g., [32]).

Acknowledgments. The authors are greatly indebted with the 81 friends and colleagues for participating in the experiment. This research was supported by Grants G.0405.08 and G.0234.08 of the Flemish Fund for Scientific Research.

References

1. Hampton, J.A.: Overextension of Conjunctive Concepts: Evidence for a Unitary Model for Concept Typicality and Class Inclusion. J. Exp. Psych.: Lear. Mem. Cog. 14, 12–32 (1988)
2. Hampton, J.A.: Disjunction of Natural Concepts. Memory & Cognition 16, 579–591 (1988)
3. Osherson, D.N., Smith, E.E.: On the Adequacy of Prototype Theory as a Theory of Concepts. Cognition 9, 35–58 (1981)
4. Hampton, J.: Conceptual Combination. In: Lamberts, K., Shanks, D. (eds.) Knowledge, Concepts, and Categories, pp. 133–159. Psychology Press, Hove (1997)
5. Zadeh, L.: Fuzzy Sets. Information & Control 8, 338–353 (1965)
6. Zadeh, L.: A Note on Prototype Theory and Fuzzy Sets. Cognition 12, 291–297 (1982)
7. Osherson, D.N., Smith, E.E.: Gradeness and Conceptual Combination. Cognition 12, 299–318 (1982)
8. Komatsu, L.K.: Recent Views on Conceptual Structure. Psych. Bull. 112, 500–526 (1992)
9. Fodor, J.: Concepts: A Potboiler. Cognition 50, 95–113 (1994)
10. Rips, L.J.: The Current Status of Research on Concept Combination. Mind and Language 10, 72–104 (1995)
11. Aerts, D.: A Possible Explanation for the Probabilities of Quantum Mechanics. J. Math. Phys. 27, 202–210 (1986)
12. Aerts, D.: The Construction of Reality and Its Influence on the Understanding of Quantum Structures. Int. J. Theor. Phys. 31, 1815–1837 (1992)

13. Aerts, D.: Quantum Structures, Separated Physical Entities and Probability. Found. Phys. 24, 1227–1259 (1994)
14. Aerts, D.: Foundations of Quantum Physics: A General Realistic and Operational Approach. Int. J. Theor. Phys. 38, 289–358 (1999)
15. Gabora, L., Aerts, D.: Contextualizing Concepts Using a Mathematical Generalization of the Quantum Formalism. J. Exp. Theor. Art. Int. 14, 327–358 (2002)
16. Aerts, D., Gabora, L.: A Theory of Concepts and Their Combinations I: The Structure of the Sets of Contexts and Properties. Kybernetes 34, 167–191 (2005)
17. Aerts, D., Gabora, L.: A Theory of Concepts and Their Combinations II: A Hilbert Space Representation. Kybernetes 34, 192–221 (2005)
18. Aerts, D., Czachor, M., D'Hooghe, B.: Towards a Quantum Evolutionary Scheme: Violating Bell's Inequalities in Language. In: Gontier, N., Van Bendegem, J.P., Aerts, D. (eds.) Evolutionary Epistemology, Language and Culture - A Non Adaptationist Systems Theoretical Approach, Springer, Dordrecht (2006)
19. Aerts, D.: Quantum Particles as Conceptual Entities. A Possible Explanatory Framework for Quantum Theory. Found. Sci. 14, 361–411 (2009)
20. Aerts, D.: Quantum Structure in Cognition. J. Math. Psych. 53, 314–348 (2009)
21. Aerts, D., Aerts, S., Gabora, L.: Experimental Evidence for Quantum Structure in Cognition. In: Bruza, P., Sofge, D., Lawless, W., van Rijsbergen, K., Klusch, M. (eds.) QI 2009. LNCS, vol. 5494, pp. 59–70. Springer, Heidelberg (2009)
22. Aerts, D., D'Hooghe, B.: Classical Logical Versus Quantum Conceptual Thought: Examples in Economy, Decision Theory and Concept Theory. In: Bruza, P., Sofge, D., Lawless, W., van Rijsbergen, K., Klusch, M. (eds.) QI 2009. LNCS, vol. 5494, pp. 128–142. Springer, Heidelberg (2009)
23. Aerts, D., D'Hooghe, B., Haven, E.: Quantum Experimental Data in Psychology and Economics. Int. J. Theor. Phys. 49, 2971–2990 (2010)
24. Aerts, D.: Quantum Interference and Superposition in Cognition: Development of a Theory for the Disjunction of Concepts (2007), Archive reference and link: http://arxiv.org/abs/0705.0975
25. Aerts, D.: General Quantum Modeling of Combining Concepts: A Quantum Field Model in Fock Space (2007), Archive reference and link: http://arxiv.org/abs/0705.1740
26. Clauser, J.F., Horne, M.A., Shimony, A., Holt, R.A.: Proposed Experiment to Test Local Hidden-Variable Theories. Phys. Rev. Lett. 23, 880–884 (1969)
27. Nelson, D.L., McEvoy, C.L.: Entangled Associative Structures and Context. In: Bruza, P.D., Lawless, W., van Rijsbergen, C.J., Sofge, D. (eds.) Proceedings of the AAAI Spring Symposium on Quantum Interaction. AAAI Press, Menlo Park (2007)
28. Bruza, P.D., Kitto, K., McEvoy, D., McEvoy, C.: Entangling Words and Meaning. In: Proceedings of the Second Quantum Interaction Symposium, pp. 118–124. Oxford University Press, Oxford (2008)
29. Bruza, P., Kitto, K., Nelson, D., McEvoy, C.: Extracting Spooky-Activation-at-a-Distance from Considerations of Entanglement. In: Bruza, P., Sofge, D., Lawless, W., van Rijsbergen, K., Klusch, M. (eds.) QI 2009. LNCS, vol. 5494, pp. 71–83. Springer, Heidelberg (2009)
30. Aerts, D.: Interpreting Quantum Particles as Conceptual Entities. Int. J. Theor. Phys. 49, 2950–2970 (2010)
31. Tsirelson, B.S.: Quantum Generalizations of Bell's Inequality. Lett. Math. Phys. 4, 93 (1980)
32. Aerts, D., Czachor, M.: Quantum Aspects of Semantic Analysis and Symbolic Artificial Intelligence. J. Phys. A-Math. Gen. 132, L123–L132 (2004)

Options for Testing
Temporal Bell Inequalities
for Mental Systems

Harald Atmanspacher[1,2] and Thomas Filk[1,3,4]

[1] Institute for Frontier Areas of Psychology, Freiburg, Germany
[2] Collegium Helveticum, ETH Zürich, Switzerland
[3] Institute of Physics, University of Freiburg, Germany
[4] Parmenides Center for the Study of Thinking, Munich, Germany

Abstract. It is shown how the concept of Bell inequalities may be used to decide whether "superposition" states exist in mental systems. For this purpose a generalized form of temporal Bell inequalities, originally developed for two-state systems, is derived for systems with any finite number of states. We propose options for testing violations of these inequalities in psychological experiments and discuss the important role of "non-invasive" measurements. Classical models can violate temporal Bell inequalitites, but observations are invasive.

Keywords: entanglement, invasiveness, neural networks, non-commutativity, temporal Bell inequalities, temporal nonlocality.

1 Introduction

From an algebraic viewpoint, the main difference between the mathematical formalism of classical physics and quantum physics is the non-commutativity of observables. In the usual framework of classical physics, observables are functions on phase space (or, more general, configuration space) with commutative pointwise product. The essential physical reason for this commutative behavior is the fundamental assumption of classical physics that observations have no influence on the state of an observed system, in particular they do not change this state.

In the mathematical formulation of quantum theory, measurements are represented by (linear, self-adjoint) operators acting on the space of states usually assumed to be a Hilbert space (i.e., essentially a vector space with a scalar product defined for vectors). This representation of observables takes into account the experimental evidence that observations (or measurements) change the state of a system. Therefore, the results of and, particularly, the resulting state after temporally successive measurements may depend on the order of the measurements. This is reflected by the non-commutativity of the mathematical representations of observables.

D. Song et al. (Eds.): QI 2011, LNCS 7052, pp. 128–137, 2011.

In the mathematical framework of quantum theory, the non-commutativity of observables is related to all other non-classical phenomena of quantum systems, such as intrinsic indeterminism, superposition states, quantum probabilities, uncertainty relations, and the violation of so-called Bell inequalities. However, non-commuting observables alone do not strictly imply these quantum features. In quantum theory, the set of observables fulfills many more conditions, and the axiomatic definition of states as expectation value functionals for observables leads to an almost fixed mathematical structure.

Recently, there have been attempts to generalize the mathematical framework of quantum theory (see [1,2]), and it is still an open question under which conditions the typical features of non-classical behavior are to be expected.

In mental systems it is obvious that observations generically influence the observed system by changing its state. Therefore, it is to be expected that observations do not commute and that a mathematical representation of these observables has to involve non-commutative structures. However, the extent to which such a non-commutative structure of mental observables leads to non-classical behavior remains open.

The "holy grail" for evident non-classical behavior would be a violation of Bell inequalities [3]. They refer to correlations between the results of measurements of different observables, and they have to be satisfied by any system for which the result of any measurement is strictly determined by the present state of the system. The assumption of such a strict determination alone is sufficient to derive Bell inequalities, not only for physical systems but for any system for which notions such as state, observable, measurement, and so on, make sense – including mental systems. A violation of Bell inequalities in mental systems would yield far-reaching insights into the nature of mental states.

In Sec. 2 we will derive a class of Bell inequalities which is particularly suited for temporal correlations between observations. Section 3 emphasizes the "non-invasiveness" of the measurements necessary to observe a possible violation of Bell-type inequalities experimentally. We will briefly describe classical models for which a non-commutative structure for observables can be defined but where a violation of Bell-type inequalities is merely the result of invasive observations. A brief summary and outlook conclude the article.

2   Bell Inequalities

In this section we will derive Bell inequalities which are particularly suited to be tested in experiments where the different observables correspond to the same type of measurements. We will first derive a conventional Bell inequality for a simple two-state system in subsection 2.1. Then, in subsection 2.2, we consider temporal Bell inequalitites for such a system, and in subsection 2.3 we discuss a more general temporal Bell inequality for an arbitrary (discrete) number of states, in particular for cases in which not all possible states are known.

Table 1. Any classical system falls into one of eight possible classes with respect to the three measurement results s_i, $i = 1, 2, 3$. Crosses under $N^-(i, j)$ mark those cases where the results s_i and s_j of measurements differ.

s_1	s_2	s_3	$N^-(1,3)$	$N^-(1,2)$	$N^-(2,3)$
+1	+1	+1			
+1	+1	−1	×		×
+1	−1	+1		×	×
+1	−1	−1	×	×	
−1	+1	+1	×	×	
−1	+1	−1		×	×
−1	−1	+1	×		×
−1	−1	−1			

2.1 Bell Inequality for a Two-State System

We assume that three observables are given with respect to which a system can be in one of two possible states characterized by + and −.[1] Later, in subsections 2.2 and 2.3, we will consider temporal versions of Bell inequalities where the measurements refer to the same experiment, but are performed successively at different moments in time.

The central assumption will be that the state of a system determines the outcome of each of the three measurements. This implies that each possible state belongs to one of eight classes, each class being labeled by the possible outcomes of the measurements (see left hand side of Table 1). It should be emphasized that we do not assume that in a particular situation we actually *know* to which class the momentary state of a system belongs. In particular, for mental systems it will be almost impossible to determine the class to which the mental state of an individual belongs. For the following arguments, it is sufficient that such an assignment of a state to one of the classes is possible in principle.

Table 1 shows that in all cases for which measurement 1 and measurement 3 yield *different* results, either measurement 1 and 2 yield different results, $s_1 \neq s_2$, or measurement 2 and 3 yield different results, $s_2 \neq s_3$. Moreover, there are also cases for which $s_1 \neq s_2$ *or* $s_2 \neq s_3$, but $s_1 = s_3$.[2] We can now deduce from Table 1 that for any given ensemble of states the following inequality holds,

$$N^-(1,3) \leq N^-(1,2) + N^-(2,3),\tag{1}$$

where $N^-(i, j)$ denotes the total number of states for which measurement i and measurement j ($i, j = 1, 2, 3$ and $i \neq j$) yield different results.

Inequality (1) is already a Bell inequality. In principle, this inequality can be tested by determining, for a given ensemble of systems (e.g., a group of individuals), the numbers $N^-(i, j)$ and then checking the results. If we assume that

---

[1] The result of a single measurement for each of the observables can also be "yes" or "no", or "0" or "1", or, more generally, a and b.

[2] The "or" here is the logical OR, not the logical "exclusive or" XOR.

the probability for being in one of the states does not depend on the particular members of the ensemble but is a general property of mental systems (with inter-individual variability), we can interpret inequality (1) as a probability relation for a particular population:

$$p^-(1,3) \leq p^-(1,2) + p^-(2,3). \tag{2}$$

This means, for each individual we only measure whether or not two of the measurements (with randomized sequence) yield different results, but we do not have to determine separately to which of the eight classes the (mental) state of each individual belongs.

It can be shown that a violation of Bell inequalities is only possible, if the three measurements do not commute. In this case quantum theory tells us that it is not possible for a state to determine the outcome with respect to each of the three measurements. (Technically speaking, there are no simultaneous eigenstates for observables which do not commute.) Therefore, the initial assumption (each state belonging to one of the eight possible classes) must be wrong. This makes the non-commutativity of observables a necessary prerequisite for a possible violation of Bell inequalities. For quantum systems this has been empirically confirmed beyond any reasonable doubt [4].

The non-commutativity of quantum observables makes it impossible to experimentally determine the precise outcomes of the corresponding measurements simultaneously. For our argumentation, however, it was important that a measurement does not change the state of a system in such a way that a second measurement yields a result different from the one it would have yielded in case the first measurement had not been performed.[3] This condition is called the "non-invasiveness" of a measurement.

In order to have non-invasive measurements (at least in a classical meaning), Bell proposed to test inequality (1) for entangled states, where only one measurement has to be performed on each subsystem. Entanglement guarantees a correlation allowing us to deduce the state of one subsystem from the measured state of the other subsystem. If the subsystems are sufficiently separated and the measurements of the two subsystems are performed almost simultaneously, the assumption of non-invasiveness is classically justified. In quantum theory, however, each of the measurements leads to a (non-local) change of the total state. Therefore, on a quantum level these measurements are invasive.

For mental systems we have no evidence for entangled states (e.g., between different individuals). Therefore, non-invasive measurements cannot be guaranteed this way. We will come back to this point in more detail in subsection 2.3.

2.2   Temporal Bell Inequalitites for Two-State Systems

Since non-commuting observables are a prerequisite for a violation of Bell's inequalities, good candidates for such observables need to be selected. Instead of

---

[3] Even if all changes were deterministic, the class to which a state belongs depends on the order in which the measurements are performed.

trying to measure two of three non-commuting observables at a time (and having to circumvent the uncertainty relations), a more suitable test for violations of Bell inequalities in mental systems is realized by so-called temporal Bell inequalities.[4] The actual measurements remain the same as discussed above, but the different observables now are measured at three different instances in time. If the temporal evolution of a system is incompatible with an observable (i.e., if the Hamiltonian does not commute with that observable), the observables corresponding to measurements at different instances may not commute. For such a situation we can reformulate the Bell inequality (1) in the following form:

$$N^-(t_1, t_3) \leq N^-(t_1, t_2) + N^-(t_2, t_3), \tag{3}$$

where now $N^-(t_i, t_j)$ refers to the number of cases where a measurement at t_i and a second measurement at t_j yield different results. The essential assumption — which generalizes the assumption that a state determines uniquely the outcome of any measurement — now is that the history of a system is fixed and that the history of the system determines the outcome of any measurement at any time.

If we assume time-translation invariance and choose the instances in such a way that $t_3 - t_1 = 2(t_2 - t_1) = 2\tau$, we obtain [7]:

$$N^-(2\tau) \leq 2N^-(\tau). \tag{4}$$

This is a sublinearity condition saying that the number of cases for which different results have been obtained with a time interval 2τ should always be smaller (or at least equal) twice the number of cases for which different results have been obtained with a time interval τ. Inequality (4) is the one we will discuss in the context of possible experiments with mental systems in order to test violations of temporal Bell inequalities.

2.3   Generalized Temporal Bell Inequalities

In this section, we will generalize the two results (1) and (4) in such a way that more than two states are permitted. This will be of relevance when we discuss the possibility of "hidden mental states", i.e. of mental states which we need not be aware of.

We assume that the sets of possible results for the three observables 1, 2, and 3 are such that it is meaningful to say that the outcomes of measurement i and j are "equal" or "different". (Technically speaking, this is e.g. realized, if the spectra of the observables are equal.) Again, the number $N^-(i, j)$ refers to the number of cases where the two observables i and j ($= 1,2$ or 3) are in different states. Then the following inequality has to hold:

$$N^-(1, 3) \leq N^-(1, 2) + N^-(2, 3). \tag{5}$$

It just follows from the fact that if the results of measurements 1 and 3 are known to be different, then either the results of measurements 1 and 2 have

---

[4] Temporal Bell inequalities were initially discussed by Leggett et al. [5] and later by Mahler [6].

to be different, or the results of measurements 2 and 3 (or both). This follows immediately from the transitivity of "being equal": If a and b are equal and b and c are equal then a and c have to be equal.

Inequality (5) is the same as for the case with only two possible measurement outcomes. Rewriting (5) as a temporal inequality and choosing the same t-values as in subsection 2.2 (and assuming time-translation invariance) we again obtain the sublinearity condition (4). Now, however, $N^-(\tau)$ refers to the number of states that are different at time t and time $t + \tau$.

Systems with an increasing number of states entail that temporal Bell inequalities are increasingly difficult to violate. However, the advantage of including more than two states is that the inequalities do not depend on the existence of "hidden states", i.e. states which one is unaware of. This will become relevant in our discussion of acategorial states below.

Note that (5) refers to a discrete number of states. The case of continuous variables is more difficult to deal with. Technically it is more difficult, because the condition of two states being equal is of measure zero. Practically it is more difficult, because the decision whether or not two states are equal is empirically more difficult to make.

3   Experimental Tests of Bell Inequalities in Mental Systems

For an experimental determination of $N^-(t_i, t_j)$ (or the corresponding probability) one might think of simply asking a subject about its mental state at time t_j and repeating the same question at time t_i. However, such observations clearly can have an influence on the mental state, so that they are typically invasive.

A similar situation occurs when individuals are asked to first memorize their mental states at time t_j and t_i and finally report whether or not the states were different. Even though in this case only one observation is made "externally", two "internal" self-observations must be made for the states to be memorized. Again, this form of self-observation may be invasive, and the later state may be different from the state which would have been realized if such a self-observation had not been made.

An example of a mental two-state system is given by the two percepts corresponding to the two possible perspectives of a Necker cube [8]. The Necker cube is a two-dimensional drawing of a cube whose perception is ambiguous with respect to its two perspectival interpretations.

Numerous publications report the distribution of dwell times for the two percepts, i.e. the probabilities $p^-(\tau)$ of perceiving different percepts at $t = 0$ and $t = \tau$). This distribution is well approximated by a gamma function (cf. Brascamp et al. [9]), which seems to indicate a violation of the temporal Bell inequality (4) in the regime of small times τ. However, these experiments are hardly non-invasive.

Even though we may never be able to fully guarantee that a measurement is non-invasive, an interesting candidate for such a measurement could be a scenario in which individuals are not asked to observe their states at t_i and t_j separately, but to judge at $t > (t_i, t_j)$ whether or not they have been the same at both instances. This would represent one single "product observation" with the possible results "same" or "different". The individual would not have to be aware of the states at time t_j and t_i separately. It is only necessary to report later whether or not they have been the same.

There is an analogue of this situation in quantum physics. In a double-slit experiment the particles may either show an interference pattern after the double slit or a single broad distribution. Any measured "which path"-information destroys the coherence between the two contributions of slit 1 and slit 2 and, therefore, the interference pattern. On the other hand, when "which path"-information is in principle unavailable, the interference patterns are observed. Similarly, the "which-state" information about the two states at time t_j and time t_i (corresponding to the "which path"-information in the double-slit scenario) destroys the non-classical behavior. Mere knowledge about whether or not the states were the same does not include "which-state" information.

As the result of such a single product measurement (same or not) depends on a correlation between states at different instances of time, we call such measurements "temporally non-local". They are also known in quantum theory, where they exhibit a somewhat tricky behavior though. Let us assume that a measurement is represented by the matrix σ_3, and the generator of the temporal evolution (the Hamiltonian H) by the matrix σ_1:

$$\sigma_3 = \begin{pmatrix} 1 & 0 \\ 0 & -1 \end{pmatrix} \quad , \quad H = g\sigma_1 = g \begin{pmatrix} 0 & 1 \\ 1 & 0 \end{pmatrix} . \tag{6}$$

Such a model was in fact investigated in the context of the bistable perception of ambiguous stimuli [10,11], and it does indeed predict a violation of Bell inequalities under certain conditions [12,7,13].

The product operator $M(t) = \sigma_3(t)\sigma_3(0)$ has eigenvalues $e^{\pm 2igt}$, where $1/g$ is the basic time scale of the evolution. These eigenvalues are not real – $M(t)$ is not a quantum observable because it is not self-adjoint. Self-adjoint combinations of $M(t)$ are the real part $S(t) = \frac{1}{2}(M(t) + M^+(t))$ with the single eigenvalue $\cos(2gt)$ and the imaginary part $A(t) = \frac{1}{2i}(M(t) - M^+(t))$ with eigenvalues $\pm \sin 2gt$.

Interpreted as the possible outcomes of single measurements, these eigenvalues already show a non-classical effect. The eigenvalues of $\sigma(0)$ and $\sigma(t)$ are ± 1 each, so their product can only be $+1$ or -1. However, the eigenvalues of the product operators assume these values only for particular values of t. This indicates that non-deterministic quantum behavior is not the result of "hidden variables". In general, temporally non-local measurements are difficult to perform for quantum systems.

In the context of mental systems such measurements may be easier. As an example for a temporally non-local measurement we mention the determination of so-called order thresholds. It has been observed [14] that for time intervals between successive stimuli that are slightly below $\approx 30 - 70\,\mathrm{ms}$, individuals are able to distinguish the stimuli as not simultaneous without being able to assign their sequence correctly. Measurements of this type are interesting candidates for product measurements with different values of time intervals τ in order to test the sublinearity condition for $p^-(\tau)$ and $p^-(2\tau)$.

Finally, we should like to mention that models with non-commuting observables can easily be found in the classical realm. Simple examples are neural networks, where the presentation of an input pattern may be interpreted as an observation and the measurement of the reaction of the network at the output nodes as the result of this observation.[5] One can easily construct examples of this type which seem to violate temporal Bell inequalities. However, the application of an input pattern is an invasive operation.

4 Conclusions

We derived an inequality for correlations between the results of observations, which can be interpreted as a temporal Bell inequality and which has to hold under the assumption that the state of a system determines the outcome of any measurement among a class of (non-commuting) observables. This temporal inequality can be tested for mental systems as well. However, the main challenge is that measurements be non-invasive, i.e., the result of a second measurement assumed to be determined by the initial state is not changed due to the first measurement. We discussed temporally non-local measurements as a possibility to circumvent the difficulties related to this challenge.

Non-commuting observables can be implemented in very simple, strictly deterministic systems, where they do not lead to a violation of Bell inequalities. Neural networks provide a simple example for *invasive* measurements which lead to a violation of Bell inequalities. Once more, this highlights the issue of invasive measurements.

The extension of Bell inequalities to systems with more than two states may prove to be relevant for non-classical mental states in the sense of acategorial states [16,17] (this term was first introduced by Jean Gebser [18]). While non-classical states of quantum systems may be interpreted as superpositions of states with classical properties, the more general notion of an acategorial state is particular suited for mental systems.

Such acategorial states may refer to transition phases between common categorial representations, and it has been proposed that they represent examples for states with "non-conceptual content" [17]. In analogy with quantum theory, any attempt to direct one's conscious attention to an acategorial state corresponds to a measurement and destroys the state. We may speculate that the decision of whether or not two successive states are the same or not, as an example for a temporally non-local, non-invasive measurement, may be suitable

---

[5] For more details see [15] where non-commutating observables were investigated for a special class of neural networks.

to provide indirect evidence for acategorial states without directly probing and thus destroying them.

Two directions of further research along the lines discussed seem to be promising: (1) performing experiments with temporally non-local, non-invasive measurements, and (2) testing Bell inequalities in recurrent and non-deterministic generalizations of neural network models.

Acknowledgement. We acknowledge discussions with P. beim Graben and H. Römer. One of us (T.F.) acknowledges partial funding of this research by the Fetzer Foundation Trust.

References

1. Atmanspacher, H., Römer, H., Walach, H.: Weak Quantum Theory: Complementarity and Entanglement in Physics and Beyond. Foundations of Physics 32, 379–406 (2002)
2. Atmanspacher, H., Filk, T., Römer, H.: Weak Quantum Theory: Formal Framework and Selected Applications. In: Adenier, G., et al. (eds.) Quantum Theory: Reconsideration of Foundations - 3, pp. 34–46. American Institute of Physics, New York (2006)
3. Bell, J.S.: On the Problem of Hidden Variables in Quantum Theory. Rev. Mod. Phys. 38, 447–452 (1966)
4. Aspect, A., Dalibard, J., Roger, G.: Experimental Test of Bell's Inequalities Using Time-Varying Analyzers. Phys. Rev. Lett. 49, 1804–1807 (1982)
5. Leggett, A.J., Garg, A.: Quantum Mechanics versus Macroscopic Realism: Is the Flux There When Nobody Looks? Phys. Rev. Lett. 54, 857–860 (1985)
6. Mahler, G.: Temporal Bell Inequalities: A Journey to the Limits of "Consistent Histories". In: Atmanspacher, H., Dalenoort, G. (eds.) Inside Versus Outside. Endo- and Exo-Concepts of Observation and Knowledge in Physics, Philosophy and Cognitive Science, pp. 195–205. Springer, Berlin (1994)
7. Atmanspacher, H., Filk, T.: A Proposed Test of Temporal Nonlocality in Bistable Perception. Journal of Mathematical Psychology 54, 314–321 (2010)
8. Necker, L.A.: Observations on Some Remarkable Phenomenon Which Occurs in Viewing a Figure of a Crystal or Geometrical Solid. The London and Edinburgh Philosophy Magazine and Journal of Science 3, 329–337 (1832)
9. Brascamp, J.W., van Ee, R., Pestman, W.R., van den Berg, A.V.: Distributions of Alternation Rates in Various Forms of Bistable Perception. Journal of Vision 5, 287–298 (2005)
10. Atmanspacher, H., Filk, T., Römer, H.: Quantum Zeno Features of Bistable Perception. Biol. Cybern. 90, 33–40 (2004)
11. Atmanspacher, H., Bach, M., Filk, T., Kornmeier, J., Römer, H.: Cognitive Time Scales in a Necker-Zeno Model for Bistable Perception. Open Cybernetics & Systemics Journal 2, 234–251 (2008)
12. Atmanspacher, H., Filk, T., Römer, H.: Complementarity in Bistable Perception. In: Atmanspacher, H., Primas, H. (eds.) Recasting Reality – Wolfgang Pauli's Philosophical Ideas in Contemporary Science, pp. 135–150. Springer, Berlin (2009)
13. Filk, T.: Non-Classical Correlations in Bistable Perception? Axiomathes 21, 221–232 (2011)

14. Pöppel, E.: A Hierarchical Model of Temporal Perception. Trends in Cognitive Science 1, 56–61 (1997)
15. Atmanspacher, H., Filk, T.: Complexity and Non-Commutativity of Learning Operations on Graphs. BioSystems 85, 84–93 (2006)
16. Atmanspacher, H.: Categoreal and Acategoreal Representation of Knowledge. Cognitive Systems 3, 259–288 (1992)
17. Feil, D., Atmanspacher, H.: Acategorial States in a Representational Theory of Mental Processes. Journal of Consciousness Studies 17(5/6), 72–101 (2010)
18. Gebser, J.: The Ever-Present Origin. Ohio University Press, Columbus (1986)

Quantum-Like Uncertain Conditionals for Text Analysis

Alvaro Francisco Huertas-Rosero[1] and C.J. van Rijsbergen[2]

[1] University of Glasgow
[2] University of Cambridge
{alvaro,keith}@dcs.gla.ac.uk

Abstract. Simple representations of documents based on the occurrences of terms are ubiquitous in areas like Information Retrieval, and also frequent in Natural Language Processing. In this work we propose a logical-probabilistic approach to the analysis of natural language text based in the concept of Uncertain Conditional, on top of a formulation of lexical measurements inspired in the theoretical concept of ideal quantum measurements. The proposed concept can be used for generating topic-specific representations of text, aiming to match in a simple way the perception of a user with a pre-established idea of what the usage of terms in the text should be. A simple example is developed with two versions of a text in two languages, showing how regularities in the use of terms are detected and easily represented.

1 Introduction

How do prior expectations/knowledge affect the way a user approaches a text, and how they drive the user's attention from one place of it to another? This is a very important but tremendously complex question; it is indeed as complex as human perception of text can be. Including such effects in the representation of text may be a relatively easy way to enhance the power of a text retrieval or processing system. In this work we will not address the question, but assume a simple answer to it, and follow it while building theoretical concepts that can constitute a tool for representing natural language text for retrieval of similar processing tasks.

The key concept to be defined will be an **Uncertain conditional** defined between lexical measurements, which will allow us to exploit structures and features from both Boolean and Quantum logics to include certain features in a text representation.

Automatic procedures for acquiring information about term usage in natural language text can be viewed as lexical measurements, and can be put as statements such as [term t appears in the text][1], to which it is possible to assign true/false values. These can be regarded as a set of **propositions**. Some

---

[1] In this paper we will use the convention that anything between square brackets [and] is a proposition.

D. Song et al. (Eds.): QI 2011, LNCS 7052, pp. 138–148, 2011.

relations between propositions have the properties of an **order relation** $\sqsubseteq$: for example, when one is a particular case of the other, e.g $P_1 =$ [term "research" appears in this text] and $P_2 =$ [term "research" appears in this text twice] we can say that $P_2 \sqsubseteq P_1$ or that P_2 is *below* P_1 according to this ordering.

The set of propositions ordered by relation $\sqsubseteq$ can be called a *lattice* when two conditions are fulfilled [2]: 1) a proposition exists that is above all the others (**supremum**), and 2) a proposition exists that is below all the others (**infimum**). When any pair of elements of a set has an order relation, the set is said to be **totally ordered**, as is the case with sets of integer, rational or real numbers and the usual order "larger or equal/smaller or equal than " $\geqslant / \leqslant$. If there are pairs that are not ordered, the set is **partially ordered**.

Two operations can be defined in a lattice: the **join** $[A \wedge B]$ is the higher element that is below A and B and the **meet** $[A \vee B]$ is the lower element that is above A and B. In this work, only lattices where both the join and the meet exist and are unique. These operations are sometimes also called conjunction and disjunction, but we will avoid these denominations, which are associated with more subtle considerations elsewhere [5].

In terms of only ordering, another concept can be defined: the **complement**. Whe referring to propositions, this can also be called **negation**. For a given proposition P, the complement is a proposition $\neg P$ such that their join is the supremum sup and their meet is the infimum inf:

$$[P \wedge \neg P = inf] \wedge [P \vee \neg P = sup] \tag{1}$$

Correspondences between two ordered sets where orderings are not altered are called **valuations**. A very useful valuation is that assigning "false" or "true" to any lattice of propositions, where {"false","true"} is made an ordered set by stating ["false" $\sqsubseteq$ "true"]. With the example it can be checked that any sensible assignation of truth to a set of propositions ordered with $\sqsubseteq$ will preserve the order. Formally, a valuation V can be defined:

$$V : \{P_i\} \rightarrow \{Q_i\}, \text{ such that } (P_i \sqsubseteq_P P_j) \Rightarrow (V(P_i) \sqsubseteq_Q V(P_j)) \tag{2}$$

where $\sqsubseteq_P$ is an order relation defined in $\{P_i\}$ and $\sqsubseteq_Q$ is an order relation defined in $\{Q_i\}$. Symbol $\Rightarrow$ represents material implication: $[X \Rightarrow Y]$ is true unless X is true and Y is false.

Another very important and useful kind of valuations is that of **probability measures**: they assign a real number between 0 and 1 to every proposition.

Valuations allow for a different way of defining the negation or complement: for a proposition P, the complement $\neg P$ is such that in any valuation V, when P is mapped to one extreme of the lattice (supremum sup or infimum inf) then $\neg P$ will be mapped to the other

$$[[V(P) = sup] \iff [V(\neg P) = inf]] \wedge [[V(\neg P) = sup] \iff [V(P) = inf]] \tag{3}$$

For Boolean algebras, this definition will be equivalent to that based on order only (1), but this is not the case for quantum sets of propositions.

A lattice and a valuation can be completed with a way to assess if a process to use some propositions to infer others is correct. The rules that have to be fulfilled by these processes are called rules of inference. In this work we do not aim to assessing the correctness of a formula, but define instead a probability measure for relations $[A\,\boxed{R}\,B]$. So we will not be exactly defining some kind of logic but using something that formally resembles it. The kind of logic this would resemble is *Quantum Logic*, which will be explained next.

1.1   Conditionals in Quantum Logics

The description of propositions about objects behaving according to Quantum Mechanics have posed a challenge for Boolean logics, and it was suggested that the logic itself should be modified to adequately deal with these propositions [19]. Von Neumann's proposal was to move from standard propositional systems that are isomorphic to the lattice of subsets of a set (distributive lattice [2]), to systems that are instead isomorphic to the lattice of subspaces of a Hilbert subspace (orthomodular lattice [1]).

A concept that is at the core of de difference between Boolean and Quantum structures is that of compatibility. Quantum propositions may be incompatible to others, which means that, by principle, they cannot be simultaneously verified. A photon, for example, can have various polarisation states, which can be measured either as linear polarisation (horizontal and vertical) or circular (left or right) but not both at a time: they are **incompatible** measurements. The propositions about a particular polarisation measure can be represented in a 2D space as two pairs of perpendicular lines $\{\{[H], [V]\}, \{[L], [R]\}\}$, as is shown in figure 1. The lattice of propositions would be completed with the whole plane [*plane*] and the point where the lines intersect [*point*]. The order relation $\sqsubseteq$ is "to be contained in", so [*point*] is contained in every line, and all the lines are contained in the [*plane*].

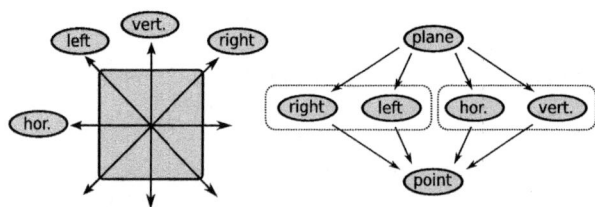

Fig. 1. System of propositions regarding the polarisation of a photon. On the left, spaces representing the propositions. On the right, order relations between them, represented with arrows. Subsets of orthogonal (mutually excluding) propositions are shown enclosed in dotted boxes.

The fact that the measurements are pairwise exclusive is not reflected in the lattice itself, but in the kind of valuations that are allowed: when $[H]$ is true, $[V]$

can only be false, but neither $[L]$ nor $[R]$ are determined. This can be described with valuation into a 3-element totally ordered set $\{false \sqsubseteq non-determined \sqsubseteq true\}$, together with two rules: 1) [only one proposition can be mapped to "true" and only one to "false"] and 2) [if one proposition from an orthogonal pair is mapped to "non-determined", the other has to be mapped to "non-determined" as well].

The rudimentary formulation of valuation rules given in the example can be, of course, improved, which can be done using a geometrical probability measure. According to Gleason's theorem [9] this probability measure can be defined by choosing a set of orthogonal directions in space with weights that sum up to 1 $\{w_i, e_i\}$, with weights that sum up to one, and computing the weighted average of the fraction of these vectors that lies within each considered subset[2], as follows:

$$V(\Pi) = \sum w_i \frac{||\Pi e_i||}{||e_i||} \tag{4}$$

The weighted orthogonal set $\{w_i, e_i\}$ is entirely equivalent to what is called **density operator** ρ and equation (4) is equivalent to the trace formula $V_\rho(\Pi) = Tr(\Pi\rho)$.

The valuations suggested in the example can be obtained by taking two of the orthogonal polarisations as e_1 and e_2 and interpreting probability 1 as "true", probability 0 as "false" and intermediates as "non-determined".

Defining conditionals in an orthomodular lattice has been a much discussed issue [8,16], and this paper does not aim to contribute to the polemic; however, we will consider two aspects of the problem from the perspective of practical applicability: the role of valuation in the definition of a logic, and the role of complement or negation.

Conditionals and the Ramsey Test Material implication $A \rightarrow B = \neg A \vee B$ is known to be problematic when requirements other than soundness are considered (like relevancy [15], context[12], etc.) and other kinds of implication are preferred in areas like Information Retrieval [17]. A key issue in the consideration of an implication is what is the interpretation of $[A \rightarrow B]$ when A is false. One possible approach to this issue is to consider "what if *it were* true", which amounts to adopting *counterfactual* conditional. If we are interested in a probability measure rather than a true/false valuation, we may as well evaluate how much imagination do we need to put into the "what if" statement: how far it is from the actual state of things. This is an informal description what is called **the Ramsey test** [7]. A simplified version of the Ramsey test can be stated as follows:

> To assess how acceptable a conditional $A \rightarrow B$ is given a state of belief, we find the least we could augment this state of belief to make antecedent A true, and then assess how acceptable the consequent B is given this augmented state of belief.

---

[2] This is not the standard formulation of the Quantum probability measure, but is entirely equivalent

In this work we will interpret *state of beliefs* as a restriction of the set of possible valuations (including probability measures) that we will use to characterise a system of propositions: in the case of a purely Quantum formulation, it would mean imposing condition on the weighted orthogonal sets. We will adopt a similar interpretation for lexical measurements in the next section.

1.2   Uncertain Conditional and Text Analysis

It has been suggested that high-level properties of natural language text such as topicality and relevance for a task can be described by means of conditional (implication) relations [18, chapter 5], giving rise to a whole branch of the area of Information Retrieval devoted to logic models [13], [20, chapter 8]. In this work we will focus on the detection of patterns in the use of words that can also be put as implication-like relations.

In this work we will focus on lexical measurements as propositions, and will adopt the concept of **Selective Eraser** (SE) as a model for lexical measurements [11]. A SE $E(t, w)$ is a transformation on text documents that preserves text surrounding the occurrences of term t within a distance of w terms, and erases the identity of tokens not falling within this distance.

A norm $|\cdot|$ for documents D is also defined, that counts the number of defined tokens (can be interpreted as remaining information). Order relations, as well as Boolean operations, can be defined for these transformations, and the resulting lattices are known to resemble those of Quantum propositions.

Order relations between SEs are defined for a set of documents $\{D_i\}$ as:

$$[E(t_1, w_1) \geqslant E(t_2, w_2)] \iff \forall D \in \{D_i\}, [E(t_1, w_1)E(t_2, w_2)D = E(t_2, w_2)D] \tag{5}$$

Since a SE erases a fraction of the terms in a document, every document defines a natural valuation for SEs on documents which is simply the count of unerased terms in a transformed document. This will be represented with vertical bars $|\cdot|$

$$V_D(E(t, w)) = |E(t, w)D| \tag{6}$$

We can also define a formula analogous to (4) defined by a set of weights and a set of documents $\{\omega_i, D_i\}$

$$V(E(t, w)) = \sum \omega_i \frac{|E(t, w)D_i|}{|D_i|} \tag{7}$$

An intuition that will guide this work is that of the point-of-view-oriented user. A user that is making a shallow reading of a text will expect only familiar terms and patterns, and will have a diminished ability to discern others that he or she does not expect. We will assume here that a topical point of view will be associated to sets of lexical propositions that are both likely and rich in order relations.

2   Conditionals for SEs

2.1   Material Implication

Using the concepts explained in the last section, we can start defining conditionals for SEs. Material implication, for example, is defined as:

$$(A \Rightarrow_m B) = (\neg A) \vee B \tag{8}$$

Two properties of probability measures can be used to evaluate a probability measure for this implication:

$$\begin{aligned} V(\neg A) &= 1 - V(A) \\ V(A \vee B) &= V(A) + V(B) - V(A \wedge B) \end{aligned} \tag{9}$$

Within a single document, the probability measure would then be:

$$V_D(E(a, w_a) \Rightarrow_m E(b, w_b)) = 1 - V(E(a, w_a)) + V(E(a, w_a) \wedge E(b, w_b)) =$$
$$= 1 - \frac{|E(a, w_a)D|}{|D|} + \min_{[E(c,w_c) \geqslant_D E(a,w_a)] \wedge [E(c,w_c) \geqslant_D E(b,w_b)]} \frac{|E(c, w_c)D|}{|D|} \tag{10}$$

This formula has all the known problems of material implication, like that of being 1 whenever $E(a, w_a)$ annihilates the document completely, so it will give probability 1 to documents without any occurrence of a or b. We have used a particular probability measure to avoid the cumbersome interpretation of what a meet and a join of SEs are. Strictly speaking, a join $E_1 \vee E_2$ would be a transformation including both E_1 and E_2. Within a single document a SE can always be found (even though it will very likely not be unique), but for a set of documents, the existence of join and meet defined in this way is not guaranteed.

2.2   Subjunctive Conditional

A much more useful probability is that of the subjunctive (Stalnaker) conditional $\Box\!\rightarrow$. The base for computing this is the Ramsey test, which starts by assuming the antecedent as true with a minimum change of beliefs. In this work we interpret that as taking the document transformed by the "antecedent" eraser $E(a, w_a)D$ as the whole document, and then compute the fraction of it that would be preserved by the further application of the "consequent" eraser $E(b, w_b)(E(a, w_a)D)$. This produces a formula resembling a conditional probability:

$$V_D(E(a, w_a) \Box\!\rightarrow E(b, w_b)) = P_D(E(b, w_b)|E(a, w_a)) = \frac{|E(a, w_a)E(b, w_b)D|}{|E(a, w_a)D|} \tag{11}$$

This number will be 1 when $E(b, w_b) \geqslant E(a, w_a)$, and will be between 0 and 1 whenever $|E(a, w_a)D| \neq 0$.

This formula still has problems when a is not in the document, because in that case both $|E(a, w_a)E(b, w_b)D| = 0$ and $|E(a, w_a)D| = 0$. A standard smoothing technique can be used in this cases using averages on a whole collection or estimates of them:

$$|E(a, w_a)E(b, w_b)\tilde{D}_0| = |E(a, w_a)E(b, w_b)D_0| + \mu|E(a, w_a)E(b, w_b)D_{avg}|$$
$$|E(a, w_a)E(b, w_b)\tilde{D}_0| = |E(a, w_a)D_0| + \mu|E(a, w_a)D_{avg}| \quad (12)$$

Conditional probability when the terms are not present in an actual document would be $\frac{|E(a,w_a)E(b,w_b)D_{avg}|}{|E(a,w_a)D_{avg}|}$. This value should be given the interpretation of "undetermined".

The final formula proposed for the probability of implication is then:

$$P_D(E(a, w_a) \,\square\!\!\rightarrow E(b, w_b)) = \frac{|E(a, w_a)E(b, w_b)D| + \mu|E(a, w_a)E(b, w_b)D_{avg}|}{|E(a, w_a)D| + \mu|E(a, w_a)D_{avg}|}$$

$$(13)$$

2.3 Topic-Specific Lattices

If we think of a user going through a text document in a hurried and shallow way, we may assume that his attention will be caught by familiar terms, and then he or she will get an idea of the vocabulary involved that is biased towards the distribution of terms around these familiar set.

Suppose we take a set of SEs with a fixed width centred in different (but semantically related) terms. We will assume that the pieces of text preserved by these can be thought as a lexical representation of the topic. In this text, we can look for order relations between narrower SEs centred in the same terms or others, as a representation of the document.

If a text is very long, or there are a large number of documents taken as a corpus to characterise lexical relations in a topic, it is not convenient to require strict conditions like $E(a, w_a)E(b, w_b)D = E(b, w_b)D$ for al large document D or for all documents D_i in a large set, because then recognised order relations would be very scarce. A more sensible approach would be to assess a probability within the text preserved by the SEs that define the topic, which would be:

$$P_{topic}(E(a, w_a) \,\square\!\!\rightarrow E(b, w_b)) =$$
$$= \max_{k_i} \left(P_{topic}([E(k_i, w_t)E(a, w_a)] \,\square\!\!\rightarrow [E(k_i, w_t)E(b, w_b)]) \right) \quad (14)$$

Restricting ourselves to the set of keywords $\{k_i\}$, the maximum value would always be for the topic-defining SE with the same central term as the antecedent SE $E(a, w_a)$ $(a = k_i)$, which simplifies the formula to:

$$P_{topic}(E(a, w_a) \,\square\!\!\rightarrow E(b, w_b)) =$$
$$= \frac{|E(a, w_a)E(b, w_b)E(a, w_t)D| + \mu|E(a, w_a)E(b, w_b)E(a, w_t)D_{avg}|}{|(E(a, w_a)D| + \mu|E(a, w_a)D_{avg}|} \quad (15)$$

for any $w_a < w_t$, where w_t is the width of the SEs used to define the topic. For large values of w_t this would be equivalent to general formula (13).

3   An Example

A particular topic might define its own particular *sub-language*; this is a well known fact, and an interesting matter for research [10]. The differences between these sub-languages and the complete language have been studied for very wide topics, such as scientific research domains [6]. In this work, we will aim to much more fine-grained topics, which could be found dominating different parts of a single document. Fine-grained sub-languages such as those would not depart from the whole language of the document significantly enough to be described grammatically or semantically as a sub-language in its own right, but will be rather a preference of some lexical relations over others.

As an illustration of how SE-based Uncertain Conditionals can be used to explore and describe the use of language characteristic of a particular, fine-grained topic, we will use two versions of a single document in different languages, and find the relations between terms chosen to define a topic. We have chosen the literary classic novel Don Quixote as the subject for examining lexical features. Two versions were used of this novel: the original in spanish [3], as it has been made available by project Gutenberg, and a translation to English by John Ormsby, obtained from the same site [4]. In this text, we define a topic by

Table 1. Characteristics of the Spanish and English version of don Quixote as a plain text sequence

language	No. of tokens	No. of terms
Spanish	387675	24144
English	433493	15714

the keywords {*sword, hand, arm, helmet, shield*} and their Spanish equivalents {*espada, mano, brazo, yelmo, adarga*} and the width for the topic-defining SEs was chosen to be 10. Co-occurrence studies have found that the most meaningful distances between terms are from 2 to 5 [14], so we took twice the highest recommended co-occurrence distance to capture also relations between terms *within* non-erased windows. Information about the text and the topics is given in table 1.

Order relations were tested with formula (15), and those implying the lower values of w_a and w_b (widths of antecedent and consequent) were taken as representative. The values can be seen in table 2.

3.1   Anomalies in the Ordering

Table 2 shows apparently paradoxical results. Relations $E(sword, 2) \square \rightarrow E(hand, 3)$ and $E(hand, 2) \square \rightarrow E(sword, 3)$, both with probabilities above 87%, do not fulfill the properties of an order relation when considered together with $E(sword, 3) \square \rightarrow E(sword, 2)$ and $E(hand, 3) \square \rightarrow E(hand, 2)$ (see figure 2). This is a result of putting together partially incompatible scenarios: $E(sword, 2) \square \rightarrow E(hand, 3)$ is evaluated in the text preserved by

Table 2. Order relations between SEs with the lower values of window width, within a topic defined by a set of erasers of width 10 centred in the same 5 words, both in their English and Spanish version. Relations $(N_1 \sqsupseteq N_2)$ represent relations $E(t_{row}, w_1) \mathbin{\square\!\!\to} E(t_{column}, w_2)$

	sword	hand	arm	helmet	shield
sword	trivial	P(2⊒3)=87%	P(1⊒3)= 93%	-	P(8⊒3) = 59%
hand	P(2⊒ 3) = 96%	trivial	P(2⊒3)= 71%	-	-
arm	P(2⊒1)=96%	P(2⊒3)=87%	trivial	P(1⊒3)=71%	P(3⊒4) = 53%
helmet	-	-	-	trivial	-
shield	P(7 ⊒3)=88%	-	P(3 ⊒3)=87%	-	trivial

	espada	mano	brazo	yelmo	adarga
espada	trivial	P(4⊒3)=67%	P(6⊒3)= 85%	-	P(2⊒7) = 52%
mano	P(2⊒ 3) = 89%	trivial	P(4⊒3)= 75%	-	P(4⊒3)= 63%
brazo	P(5⊒3)=89%	P(3⊒3)=94%	trivial	-	P(1⊒3) = 74%
yelmo	-	-	-	trivial	-
adarga	P(6 ⊒3)=94%	P(3 ⊒3)=94%	-	-	trivial

$E(sword, 10)$ and $E(hand, 2) \mathbin{\square\!\!\to} E(sword, 3)$ is evaluated in the text preserved by $E(hand, 10)$.

Anomalies in the order can be resolved by simply choosing some of the relations on the basis of their higher probability (in this case, $E(hand, 2) \sqsupseteq E(sword, 3)$ with 96% over $E(sword, 2) \mathbin{\square\!\!\to} E(hand, 3)$ wiwth 87%, or collapsing the involved SEs in a class of equivalence, so the inconsistency is removed.

3.2 Lattices for Two Languages

The sets of relations obtained are strikingly similar for the two languages, with more differences polysemic terms like "arm" (which appears in spanish with different terms for its noun meaning and for its verb meaning) and "sword" which corresponds to different kinds of weapons with their own name in Spanish, from which "sword" is just the most frequent. Moreover, the anomaly in the orderings of SEs centred in "sword" and "hand" does not appear between their spanish counterparts "espada" and "mano", but is replaced by a very similar pair of relations.

This kind of analysis provides a promising way of finding regularities between different languages, or even analogies between different terms in the same language. It is easy to isolate the transformations needed to go from the English

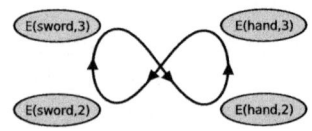

Fig. 2. Anomalous ordering of four SEs in the English topical lattice

Lattice to the Spanish one, as a lattice morphism. The differences of both could even suggest a valuation, a mapping to a simpler common lattice.

4   Discussion and Conclusion

In this work, we have shown how the framework of SEs provides a natural platform to define logical relations resembling those employed in Boolean logics, but also more complex ones, like the subjunctive conditional. Quantitative implementation follows naturally from the parallel between lexical measurements and quantum ideal measurements, producing a formula that is both simple and easy to compute for concrete cases.

The proposed formula also allows to include relations restricted to only a chosen bit of the text, that surrounding the occurrences of keywords. This allows to extract relations between terms that can be expected to be characteristic of the text about a particular topic.

The proposed formula was applied to a simple example, with very interesting results. Two main features can be observed in the results:

1. Anomalies can appear in the resulting order relation, coming from the existence of transformations that are incompatible in the sense of quantum incompatibility. These can be removed easily if a proper lattice-valued representation is to be obtained, but can also be studied as an evidence of useful patterns as well.
2. The relation structures between SEs make visible common features of the representation of a text in different languages: terms that mean something similar will be embedded into similar patterns of relations.

As a matter for future research, both observations can be explored further: the causes and characteristics of the anomalies in order relations between SEs as assessed by uncertain conditionals, and the possibility of putting the multi-language representation in terms of morphisms between lattices of SEs.

In particular, having similar lattices for two versions of the same text in different languages invites to find an optimal way of defining a common valuation that would assign both lattices to a simpler third lattices with their common features. This, in particular, is a very promising direction of research, and a novel approach to multi-lingual text processing.

Acknowledgements. This work was developed under the funding of the Renaissance Project "Towards Context-Sensitive Information Retrieval Based on Quantum Theory: With Applications to Cross-Media Search and Structured Document Access" EPSRC EP/F014384/1, School of Computing Science of the University of Glasgow and Yahoo! (funds managed by Prof. Mounia Lalmas).

References

1. Beltrametti, E.G., Cassinelli, G.: The logic of Quantum Mechanics. Addison Wesley, Reading (1981)

2. Burris, S., Sankappanavar, H.P.: A Course on Universal Algebra. Springer, Heidelberg (1981)
3. de Cervantes-Saavedra, M.: Don Quijote. Project Gutemberg (1615)
4. de Cervantes-Saavedra, M.: The ingenious gentleman Don Quixote of La Mancha. Project Gutemberg, translation by John Ormsby (1885) edition, 1615
5. D'Hooghe, B., Pyackz, J.: On some new operations on orthomodular lattices. International Journal of Theoretical Physics 39(3), 641–652 (2000)
6. Friedman, C., Kra, P., Rzhetsky, A.: Two biomedical sublanguages: a description based on the theories of Zellig Harris. Journal of Biomedical Informatics 35(4), 222–235 (2002)
7. Gärdenfors, P.: Belief revisions and the Ramsey test for conditionals. The Philosophical Review 95(1), 81–93 (1986)
8. Gardner, M.: Is Quantum Logic Really Logic? Philosophy of Science 38(4), 508–529 (1971)
9. Gleason, A.M.: Measures of the closed subspaces of the hilbert space. Journal of Mathematics and Mechanics 6, 885–893 (1957)
10. Harris, Z.: Discourse and Sublanguage, ch. 11, pp. 231–236. de Gruyter, Berlin (1982)
11. Huertas-Rosero, A., Azzopardi, L., van Rijsbergen, C.: Characterising through erasing: A theoretical framework for representing documents inspired by quantum theory. In: vam Rijsbergen, C.J., Bruza, P.D., Lawless, W. (eds.) Proc. 2nd AAAI Quantum Interaction Symposium, pp. 160–163. College Publications, Oxford (2008)
12. Huibers, T., Bruza, P.: Situations, a general framework for studying information retrieval. Information Retrieval: New Systems and Current Research 2 (1994)
13. Lalmas, M.: Logical models in information retrieval: Introduction and overview. Information Processing and Management 34(1), 19–33 (1998)
14. Lund, K., Burgess, C.: Producing high-dimensional semantic spaces from lexical cooccurrence. Behavior Research Methods, Instruments and Computers 28(2), 203–208 (1996)
15. Mares, E.D.: Relevant Logic: A Physlosophical Interpretations. Cambridge University Press, Cambridge (2004); Preface: The author claims that this kind of logic is suitable for dealing with semantics Introductions: Non Sequitur is bad
16. Pavicic, M., Megill, N.: Is Quantum Logic a Logic?, ch. 2, pp. 23–47. Elsevier, Amsterdam (2004)
17. van Rijsbergen, C.J.: A new theoretical framework for information retrieval. In: SIGIR 1986: Proceedings of the 9th Annual International ACM SIGIR Conference on Research and Development in Information Retrieval, pp. 194–200. ACM, New York (1986)
18. van Rijsbergen, C.J.: The Geometry of Information Retrieval. Cambridge University Press, Cambridge (2004)
19. von Neumann, J., Birkhoff, G.: The logic of quantum mechanics. Annals of Mathematics 43, 298–331 (1936)
20. Widdows, D., Kanerva, P.: Geometry and Meaning. Cambridge University Press, Cambridge (2004)

Modelling the Acitivation of Words in Human Memory: The Spreading Activation, Spooky-activation-at-a-distance and the Entanglement Models Compared

David Galea[1], Peter Bruza[1], Kirsty Kitto[1], Douglas Nelson[2], Cathy McEvoy[2]

[1] Queensland University of Technology, Brisbane, Australia
[2] University of South Florida, Tampa, USA

Abstract. Modelling how a word is activated in human memory is an important requirement for determining the probability of recall of a word in an extra-list cueing experiment. The spreading activation, spooky-action-at-a-distance and entanglement models have all been used to model the activation of a word. Recently a hypothesis was put forward that the mean activation levels of the respective models are as follows:

Spreading ≤ Entanglment ≤ Spooking-action-at-a-distance

This article investigates this hypothesis by means of a substantial empirical analysis of each model using the University of South Florida word association, rhyme and word norms.

1 Introduction

In extra-list cuing, participants typically study a list of to-be-recalled target words shown on a monitor for 3 seconds each (e.g., planet). The study instructions ask them to read each word aloud when it appears and to remember as many as possible, but participants are not told how they will be tested until the last word is shown. The test instructions indicate that new words, the test cues, will be shown and that each test cue (e.g., universe) is related to one of the target words just studied. These cues are not present during study (hence, the name Extra-list cuing). As each cue is shown, participants attempt to recall its associatively related word from the study list.

A crucial aspect of producing models that predict the probability of recall is modelling the activation of a target word in memory prior to cuing. Much evidence shows that for any individual seeing or hearing a word activates words related to it through prior learning. Seeing planet activates the associates earth, moon, and so on, because planet-earth, planet-moon, moon-space and other associations have been acquired in the past. This activation aids comprehension, is implicit, and provides rapid, synchronous access to associated words. Therefore, some models of activation fundamentally rely on the probabilities of such associations.

D. Song et al. (Eds.): QI 2011, LNCS 7052, pp. 149–160, 2011.

Recently, three activation were compared:[1] Spreading activation, Spooky-action-at-a-distance and a model inspired by quantum entanglement. It was hypothesized that the the spreading activation model underestimates the activation level of a target, whereas the Spooky-action-at-a-distance model may overestimate it. In short this hypothesis places the three models in relation to their mean levels of activation as follows:

$$\text{Spreading} \leq \text{Entanglment} \leq \text{Spooking-action-at-a-distance}$$

Here, we investigate the correctness of this hypothesis with a substantial empirical analysis utilising the University of South Florida word association, rhyme and word fragment norms [4]. We begin by describing how each of the models accounts for activation.

2   Activation Models

In order to aid in understanding the implementation of the three models consider the following situation in which there is a hypothetical target with two associates, a single associate-to-target and an associate-to-associate links.
For computational purposes, the above network may be represented using the following matrix,

2.1   Spooky Action at a Distance

The Spooky Action at a Distance Model is computed via the following formula:

$$S(T) = \sum_i S_{T,i} + \sum_i S_{i,T} + \sum_i \sum_j S_{i,j} \tag{1}$$

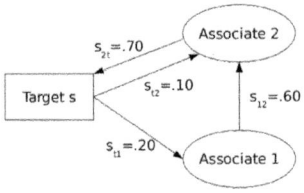

Fig. 1. A hypothetical target with two associates and single associate-to-target and associate-to-associate links [3]

Table 1. Matrix corresponding to hypothetical target shown in Fig. 1. Free assoication probabilities are obtained by finding the row of interest(the cue) and running across to the associate word obtained [2].

	Target (t)	Associate 1 (a_1)	Associate 2 (a_2)
Target (t)		0.2	0.1
Associate 1 (a_1)			0.6
Associate 2 (a_2)	0.7		

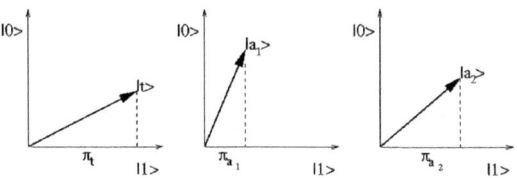

Fig. 2. Three bodied quantum system of words [1]

Where,

$$S_{i,T} = Pr(\text{Word}_i \mid T) \ , \ \ S_{T,i} = Pr(\text{Word}_i | T) \ , \ \ S_{i,j} = Pr(\text{Word}_i \mid \text{Word}_j) \ \ (2)$$

And,

$$\text{Word}_{i,j} \in \text{Target Associates}$$

Noting that $S_{i,T}$, $S_{T,i}$ and $S_{i,j}$ represent free association probabilities, i.e. $S_{i,j} = Pr(\text{Word}_i \mid \text{Word}_j)$ represents the probability that Word_i is produced when Word_j is used as cue in free association experiments [1]. Taking the example from Fig.1,

$$S(T) = (0.1 + 0.2) + (0 + 0.7) + (0.6 + 0) = 1.6.$$

2.2   Spreading Activation Model

The Spreading Activation Model is computed via the following formula:

$$S(T) = \sum_i S_{T,i}\, S_{i,T} + \sum_i \sum_j S_{T,i}\, S_{i,j}\, S_{j,T} \tag{3}$$

Where $S_{i,T}$, $S_{T,i}$ and $S_{i,j}$ are defined in the same manner as for the Spooky Action at a Distance model [1]. Taking the example from Fig.1,

$$S(T) = (0.07) + (0.2)(0.6)(0.7) + (0.1)(0)(0) = 0.014$$

2.3   Entanglment Activation Model

An alternative way to model activation is to view a targets network as a composite quantum system. Using the example of Fig. 1 to view a targets association network, this would translate into a quantum system modelled by three qubits. Fig. 2 depicts this system, where each word is in a superposed state of being activated (denoted by the basis state $|1\rangle$) or not activated (denoted by the basis state $|0\rangle$).

Thus the states of the words in the associative network are represented as,

$$|t\rangle = \bar{\pi}_t \, |0\rangle + \pi_t \, |1\rangle \ , \ \ |a_1\rangle = \bar{\pi}_{a_1} \, |0\rangle + \pi_{a_1} \, |1\rangle \ , \ \ |a_2\rangle = \bar{\pi}_{a_2} \, |0\rangle + \pi_{a_2} \, |1\rangle \tag{4}$$

While the amplitudes of the respective qubits can be derived from the matrix depicted in Table 1. Consider the column associate a_2. The two non-zero values in this column represent the level and the number of times associate a_2 is recalled in a free association experiment. Intuitively, the more non-zero entries and the higher the values, the more a_2 is activated. One way to formalize this is to take the square root of the average of these values as being the amplitude. For example $\pi_{a_2} = \sqrt{0.35}$.

The state ψ_t of the most general combined quantum system is given by the the tensor product of the individual states,

$$\psi_t = |t\rangle \otimes |a_1\rangle \otimes |a_2\rangle, \tag{5}$$

The intuition behind entanglement activation is the target t activates its associative structure in synchrony [1]. This is modelled using an entangled state, state.

$$\psi_t' = \sqrt{p_0}\,|000\rangle + \sqrt{p_1}\,|111\rangle, \tag{6}$$

which represents a situation in which the entire associative structure is either completely activated ($|111\rangle$) or not activated at all ($|000\rangle$). The entanglement model is fundamentally different to the spreading activation and the spooky-action-at-a-distance model as it models the target and its associative network as a non-separable structure. Formally, the state represented in Eq. 7 cannot factorise into states corresponding to individual words in the network and cannot be written in the form of Eq. 6.

The question remains how to ascribe values to the probabilities p_0 and p_1. In QT these values would be determined by the unitary dynamics evolving ψ_t into ψ_t', however no such dynamics exist for modelling the states of words in human memory. One approach is to assume the lack of activation of the target is determined solely in terms of lack of recall of any of the associates [2], that is,

$$p_0 = \left(1 - Pr(\bar{T})\right)\left(1 - Pr(\bar{a}_1)\right)\left(1 - Pr(\bar{a}_2)\right) \tag{7}$$

$$p_1 = 1 - p_0 = 1 - \left(1 - Pr(\bar{T})\right)\left(1 - Pr(\bar{a}_1)\right)\left(1 - Pr(\bar{a}_2)\right) \tag{8}$$

Given that p_1 refers to the probability of the target being activated, this reflects the strength of activation, namely $S(T)$. Using (15) as a basis we can easily extrapolate the model to generalise a set of rules to model a network of a Target T with a set of Associates [1]:

$$S(T) = 1 - \prod_i \left(1 - \overline{Pr(\text{Word}_i)}\right). \tag{9}$$

$$\overline{Pr(\text{Word}_i)} = \frac{1}{m_T} \sum_j Pr(\text{Word}_i \mid \text{Word}_j). \tag{10}$$

$$m_T = \{Pr(\text{Word}_i \mid \text{Word}_j) \mid \text{Word}_j \neq 0\}. \tag{11}$$

$$\text{Word}_k \in \text{Target Associates} + \text{Target}. \tag{12}$$

Table 2. Matrix corresponding to hypothetical target shown in Fig. 1. Free assoication probabilities are obtained by finding the row of interest(the cue) and running across to the associate word obtained [2].

	Target (t)	Associate 1 (a_1)	Associate 2 (a_2)
Target (t)		0.2	0.1
Associate 1 (a_1)			0.6
Associate 2 (a_2)	0.7		
$Pr(\text{Word}_i)$	0.7	0.2	0.35

Taking the example from Fig.1,

$$S(T) = 1 - (1 - 0.7)(1 - 0.2)(1 - 0.35) = 0.844$$

3   Analysis of Activation Models

Given that the focus of this paper lies on modelling the activation for each of the three models and evaluating their performance against one another, two sets of analysis were performed.

The first was centred on analysing each model individually, and in doing so, the distribution of the results was assessed on whether they exhibited normal like distributions. A key feature of normality is that it allows for the standard measure of centrality, i.e. the mean, median and mode to be used as the central platform coupled with the standard deviation to aid in understand the distribution of the results. To accompany that, a similar yet simpler analysis was performed on the errors of activation vs. the probability of recall. The purpose of which was to again seek a normal like distribution to justify the use of the mean as a potential characteristic for comparison, but furthermore to gain an understanding as to how the model compared to the observed data and in doing so to gain a better understand on how it performed overall.

The second area of analysis involved assessing the original conjecture regarding the relative performance of the three models. The mean was chosen as the figure for comparison pending all the three models fitted values could be definitively shown to follow a Normal Distribution.

The University of South Florida supplied the data set used for the testing, which was comprised of 4068 individual test cases[4]. In the analysis to follow activation levels were computed for each target and an error analysis performed against the probability of recall. The cue process is ignored in this analysis in order to focus on activation.

3.1   Spooky Action at a Distance

The Spooky Action at a Distance Activation was computed against all test cases produced the following results:

Table 3. Descriptive Statistics on Spooky-Activation-at-a-Distance

	Target Activation
Mean	0.327203
Median	0.303077
Mode	0.43
Standard Deviation	0.143161
Range	1.6775
Minimum	0.0525
Maximum	1.73

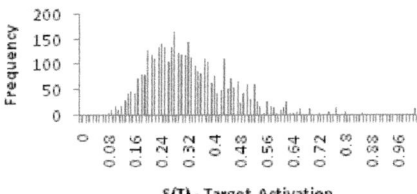

Fig. 3. Histogram of Spooky-Activation-at-a-Distance Activation Recall

Here we observe that on average the activation is fairly low (Mean = 0.327), coupled with an almost matching median and low standard deviation is it fair to suppose that its distribution would be fairly centred, dense and akin to that of a true Normal Distribution. The maximum value of 1.73 is greater than 1, as unlike spreading activation; the activation level for this model is not a probability. However, as values greater than 1 were rarely observed, these were treated as flaws/outliers for the purposes of this analysis and the spooky activation modelled was thereby assumed to generate a probability of recall. The histogram of activation levels is depicted in Fig. 3.

From the histogram it is evident that the activations are in fact robustly Normally Distributed ($N(0.327, 0.02)$). As stated previously given the low standard deviation this allows a permissible basis to establish a profile of the model based on the mean and furthermore its use as figure for comparison. To reinforce this, a further investigation was made into measuring the Target Activation against the Fitted Probability of Recall, the Results of which are shown in Fig. 4.

Fig.4 indicates that there is strong evidence that the errors are Normally Distributed, and from which the original proposition to use the Mean (-0.21961) as a basis is supported. These results show great promise for development. The under-fitting of the probability of recall is to be expected in a good model as the cue process is not present to supplement the activation levels.

3.2 Spreading Activation Model

The Spreading Activation Equation was computed against the same test cases and produced the following results:

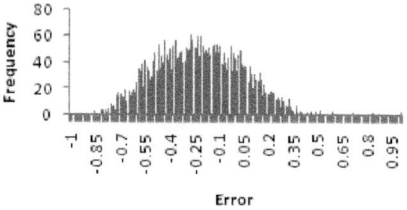

Fig. 4. Histogram of Spooky-Activation-at-a-Distance Activation Recall vs Probabi3ity of Recall ($\sigma = 0.267934$)

Table 4. Descriptive Statistics on Spreading Activation

	Target Activation
Mean	0.009919
Median	0.003736
Mode	0
Standard Deviation	0.019087
Range	0.363667
Minimum	0
Maximum	0.363667

Here we observe that on average the activation is extremely low (Mean = 0.009919), coupled with an almost matching median and particularly low standard deviation which implies that it would be fair to conclude that its distribution would be analogous to that of a Normal Distribution. In order to gain a better perspective into the distribution of the Activations, a histogram was generated as shown below,

From the histogram it is evident that the Activations are only loosely Normally Distributed $N(0.009919, 0.00001)$. The tailing right complementing the relative high upper maximum 0.363667 makes the claim of Normality hard to justify. In order to validate this, an investigation into the target activation against the probability of recall (as with the Spooky at a Distance Model) was performed. The results of which are shown in Fig. 6.

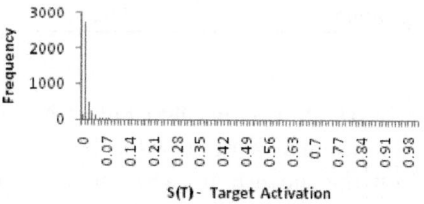

Fig. 5. Histogram of Spreading Activation Recall

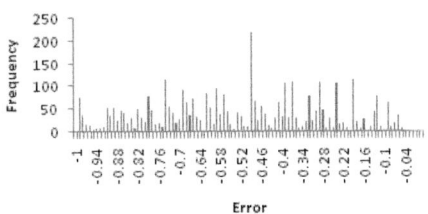

Fig. 6. Histogram of Spreading Activation Recall vs Probability of Recall ($\sigma =$ 0.248759)

It is clearly evident through the random nature of the distribution of the errors that no relationship exists (Mean Error $= -0.52973$). As a result we infer that the inclusion of the Cue into the activation procedure does not provide any insight into its ability to accurately activate target across any spectrum. We conclude that the Spreading Activation model is likely to be a poor estimator.

3.3   Entanglement Activation Model

The entanglement activation model was computed against all test cases and produced the following results as shown in Table 5.

Here we observe that on average the activation is quite high (Mean $= 0.668155$), coupled with an almost identical median and principally low standard deviation (relative to the mean) it would be fair to speculate that its distribution would be comparable to that of a dense normal Distribution. The distribution of activations is shown in Fig. 7.

Table 5. Descriptive Statistics on Entanglement Activation

	Target Activation
Mean	0.668155
Median	0.670558
Mode	0.867404
Standard Deviation	0.094696
Range	0.622444
Minimum	0.340501
Maximum	0.962944

The activations are robustly Normally Distributed $N(0.668155, 0.009)$. Consequently we identify that there is a permissible basis to establish an overview of the model centred on the mean and enable it as figure for comparison. To reinforce this, a further investigation was made into measuring the Target Activation against the Fitted Probability of Recall. The results of which are shown in Fig. 8.

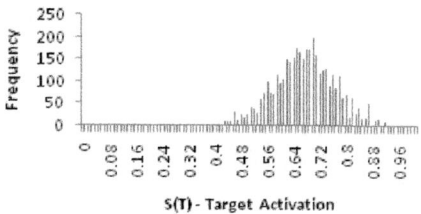

Fig. 7. Histogram of Entanglement Activation

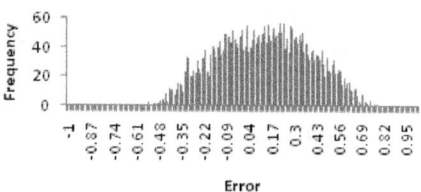

Fig. 8. Histogram of Spooky-Activation-at-a-Distance Activation Recall vs Probability of Recall ($\sigma = 0.267986$)

We observe that the errors are strongly Normally Distributed, and from this we conclude that the original proposition to use the mean (0.121345) as a basis is supported. In addition, the activations clearly overfit the probability of recall even without the cue process being considered. This propensity to overfit is something that must be closely monitored for further development as given the cue process is missing, traditionally we should expect lower activation results.

4   Discussion

The primary focus of this paper was to investigate the conjecture:

$$\text{Spreading Activation} \leq \text{Entanglement} \leq \text{Spooky}$$

Whilst the Spreading Activation Model was found to be unstable, imposing instability as an inherent feature of the model the previous conjecture simply becomes a test of whether the following relation holds

$$\text{Spreading Activation} \leq \text{Entanglement} \leq \text{Spooky}$$

Given the respective averages are Spreading = 0.009919 , Entanglement = 0.668155 , Spooky = 0.327203, The relations,

$$\text{Average Spreading} \leq \text{Average Entanglement, Average Spooky}$$

are upheld, however the following does not hold when tested upon the empirical data.

$$\text{Average Entanglement} \leq \text{Average Spooky}$$

The spreading activation model is unstable and not conducive to any generalisations. The analysis does support the view in the literature that it considerably underestimates the activation level. The entanglement activation model considerably overestimates the level of activation. The cause is the nave assumption behind Eq. 10 and 11. The strongly normal character of the of the activation distribution suggest that the bias can be corrected via a single scaling parameter applied to the probability component of Eq. 11. Alternatively, it may be handled via the introduction of an error term. Development of both adjustments to the current model is the subject of further research. The resulting model is not likely to be a better activation model than spooky because of both models have almost identical standard deviations on their errors with respect to probability of recall.

5   Summary and Outlook

The aim of this article was a detailed analysis of three models of target word activation in human memory: the spreading activation, the spooking-action-at-a-distance model and the entanglement model. Previous research has hypothesised that the mean levels of activation would be:

$$\text{Spreading Activation} \leq \text{Entanglement} \leq \text{Spooky}$$

However, the analysis presented in this paper revealed that:

$$\text{Spreading Activation} \leq \text{Spooky} \leq \text{Entanglement}$$

It was found that the spreading activation is unstable. Both the spooky and the entanglement activation models are normally distributed with respect to the error against the probability of recall which bodes well for future development of these models. The entanglement activation model overestimates the activation level, however the prospects to use simple means to mitigate the bias are good.

Clearly, the entanglement model is exhibiting great potential as a model of activation. Given that the model is still in its primitive stages of development and that there is considerable uncertainty in forming the dynamics of the entangled system (π_1 , π_2), we identify that reworking the foundations of these dynamics would prove highly difficult, and consequently further research will focus on the examing three different scenarios:

1. Develop a formalised structure for the existing activation formula and modifying it to increase performance.

   Currently $S(T)$ takes the form,

$$S(T) = 1 - f(T , A) ; A = \{A_i \,|\, A_i = \text{Associate } i \text{ to Target } T\} . \qquad (13)$$

Where,

$$f(T, A) = 1 - \prod_i \left(1 - \overline{Pr(\text{Word}_i)} \right). \tag{14}$$

It can be easily shown that this function lies in the range $[0, 1]$. Its current form thereby has a greater tendency for $f(T, A) \to 0$ as the number of associates increases. One way to overcome this would be to weight the Associates probabilities according to their strength in their respective word association networks. Consequently $f(T, A)$ would take the form,

$$f(T, A) = 1 - \prod_i \left(1 - W(A_i)\overline{Pr(\text{Word}_i)} \right). \tag{15}$$

Where $W(A_i)$ is the weighted scalar for the associate probability. This adjustment will also be designed to take the current issues with associate Probability calculation.

2. Investigate patterns that may exist in the word networks and adjusting the formulae for $S(T)$ to accomodate for each scenario.

Whilst the average was chosen as the most approprite measure of comparison between the three models due to the normal-like distribution that each exhibited, there were many cases in which the original proposition held. The violations found that word networks exhibiting certain trends satisfied the constraints whilst others didn't. Consequently, it appears that the structure of the word association network plays a great role in its respecitive activation level. At present, word association structure is currently being examined in detail to identify firstly whether a set of network ypes exists and from which how the current model for activation should be altered to accomodate each type.

3. Develop a unitary transformation U which transforms the product state ψ_t (equation (5)) into the entangled state ψ_t' (equation (6)). Quantum computing offers some potentially useful transformations which may be investigated for this purpose.

Following on from the previous ideology, if the influence on the word assoication network shows that its contribution and inclusion is not yielding better results a complete rework of the fundamental probabalistic formulation for $S(T)$ will be developed were the naive assumption being that the Target activates its Associates in synchrony will be challenged so that more sophisticated models can be developed.

References

1. Bruza, P.D., Kitto, K., Nelson, D., McEvoy, C.: Extracting spooky-activation-at-a-distance from considerations of entanglement. In: Bruza, P., Sofge, D., Lawless, W., van Rijsbergen, K., Klusch, M. (eds.) QI 2009. LNCS, vol. 5494, pp. 71–83. Springer, Heidelberg (2009)

2. Bruza, P., Kitto, K., Nelson, D., McEvoy, C.: Is there something quantum-like about the human mental lexicon? Journal of Mathematical Psychology 53, 362–377 (2009)
3. Nelson, D., McEvoy, C., Pointer, L.: Spreading activation or spooky activation at a distance? Journal of Experimental Psychology: Learning, Memory and Cognition 29(1), 42–52 (2003)
4. Nelson, D., McEvoy, C., Schreiber, T.: The University of South Florida, word association, rhyme and word fragment norms. Behaviour Research Methods, Instruments and Computers 36, 408–420 (2004)

Senses in Which Quantum Theory Is an Analogy for Information Retrieval and Science

Sachi Arafat

University of Glasgow

Abstract. Quantum theory (QT) addresses phenomena of the physical world, information retrieval (IR) is concerned with social and psychological phenomena in world of human-created objects. Recent approaches to IR employ the mathematics of QT, but in doing so, inherit notions particular to physical phenomena such as entanglement and superposition; and is lead to wonder about laws of dynamics, and general scientific concepts such as invariance. QTIR contends that their use of QT is analogical. In what senses is this the case? This paper explores the senses of this analogy and the way IR is (thereby) 'inspired' by QT.

1 Introduction

Quantum-theory inspired information retrieval (QTIR)[1] is based on *scientific loans* from QT. Scientific loans involve *"the development and reworking of cognitive material [ideas] that pre-exists .. [scientific theories] .., necessitating the creative employment of ideas from adjacent fields, Bachelard's 'scientific loans'* [2, p37] and are a common occurrence. IR and QT are not at all adjacent fields in the traditional sense since the objects of study, the social and natural worlds, are radically different; and IR (primarily) is the techné end of information science (IS) which P. Wilson, an IS pioneer, says is "a fascinating combination of engineering [the IR side], an odd kind of materials science, and social epistemology" [4] where 'materials science' corresponds to dealing with artefacts (e.g. documents). Yet, it is contended in [5] and following works that QT is relevant for IR, and possibly (by extension) for information science in general. Its relevance is based initially on the tacit assumption that there can be "creative employment" of ideas from QT through which to recast phenomena particular to IR, and that this could lead to useful insights for retrieval. Mathematical methods from QT: the representation of states (of systems) and state-change; the ability to coherently integrate uncertainty through probability, (spatial) relationships through the Hilbert space, accurate rational expression of phenomena through a corresponding logic on the Hilbert space; serve to indicate three anticipated benefits for IR: (1) a similar (possible) integration

---

[1] As opposed to "quantum-theoretic IR", since it is not quite clear in what senses IR can claim to be so, and this depends on the coherence of analogies.

D. Song et al. (Eds.): QI 2011, LNCS 7052, pp. 161–171, 2011.

of formal approaches to IR, (2) the possibility of adequately modelling complex phenomena in the psycho-social sphere, and (thereby) (3) the possibility of acquiring a "more scientific" status, at least to the extent of accommodating vertical development in IR research to complement statistics-mediated ("ad-hoc") development through experimentation. The mathematics of QT, however, was developed with regard to physical phenomena; thus, even in mathematically borrowing from QT, one inevitably needs to address, for the sake of appropriately using/interpreting the mathematics, how the physical phenomena would correspond to phenomena in IR. And since the quantum-world is ontologically different from the psych-social world, the primary mode of relation is analogy.

QTIR research also (perhaps inadvertently) leads to borrowing from QT its scientific method of expression, i.e. (1) the way physical phenomena is represented by mathematical structures and then (2) the kind of statements made using these structures, e.g. the mapping of physical properties to subspaces in the modelling experimental phenomena from *yes-no* experiments and the subsequent "descriptive mathematical" employment of corresponding symbols in elucidating other phenomena.

Although there are no rules as such for how one 'ought to be inspired by QT', all three anticipated benefits of using QT for IR, if a rational procedure for their procurement is to be sought, seem to depend on adequate development of analogical relationships between QT and IR. These relationships are at the level of: (1) ontology and epistemology, (2) phenomena, (3) modelling (by mathematical representation) and expression. The first level concerns objects and agents (such as researcher, user, 'system', or nature) in the corresponding worlds of QT or IR, the modes of interaction between agents/objects, and the way agents/objects appear to other agents. The second category refers to intuitive understandings of particular regularities (phenomena) 'within' objects and between objects and agents. The third category refers to effective representation, construction of propositions, and formal objectification and analysis of the intuitively understood regularities in the prior level. A comprehensive exploration of these levels is indeed necessary, and an initial attempt can be found in [1], parts of which are cited below. However, this would require a much larger space than afforded here, so I opt instead to motivate such an investigation by presenting some research questions that manifest at each level, usually following a brief discussion of the research concerns at that level. I devote more space to the first level since the other levels are founded upon it, and thereby to set the scene. The first such question in this regard is (Q1) "do these three categories of analogy correspond to three separate (legitimate) aspects of IR research, is that how IR should be categorised?" The general aim of this paper is to initiate the construction of a discursive framework through elucidation of basic questions. It can be seen as a background discussion, and 'discursive complex', to several of the technical works in QTIR and the issues have surfaced therein (see references in [6]).

2   Analogy at the Level of Ontology and Epistemology

The basic picture of agents in Fig. 2 shows some obvious differences between
IR and QT, from the theoretician's (T) point of view[2] IR has two modes: the
retrieval experiment (RE) that contains one more agent than the quantum experiment scenario (Q), and the retrieval scenario (RS) which has no O (not
shown). In RE, the primary agent is O as it defines the purpose of the scenario
as experiment, from which knowledge would be gathered that O makes use of. In
RS, the primary agent is the user, as the purpose of the scenario is to fulfil their
information need. In Q, it is O since the purpose there is to gather knowledge
of N. The goal of O in Q is to understand N, an object of a different type than
itself, about which it knows nothing without experiment.

In contrast, the goal of O in RE is to understand U through U-S-C interactions,
but it already knows C to a maximal extent (usually due to having designed it)
while being able to "put itself in the shoes of"/empathise with U, since O and
U are of the same type. The goal for U in RS/RE and O in Q is to understand
something about C and N through the means of S and M respectively. The
notation X-Y-Z/P should be read as the perspective of X observing Z or P
through the means of Y - there is further elaboration of these views/perspectives
in [1, ch3]. In the proceeding, I first discuss intrinsic differences between C and N
(2.1), between knowing/observing C and N through S and M (2.2), what aspects
therein are available for modelling for QTIR and what type of inquiries could
be accommodated by such models (2.3).

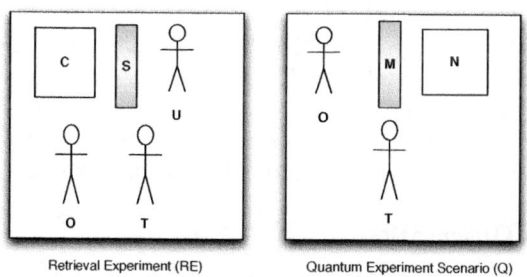

Retrieval Experiment (RE)   Quantum Experiment Scenario (Q)

Fig. 1. Basic Pictures; T=theoretician (us), O=researcher/experimental observer,
U=user, N=nature, M=measuring devices, S=system, C=collection

2.1   Nature at Micro-level vs Collection

C traditionally refers to a set of documents, abstract objects "out there" to be
"brought-near" or retrieved. A document may be an image, and in the context

---

[2] By analysing scenarios hypothetically we take the role of a theoretician. A theoretician's purpose is to assess the conditions of possibility of interactions between agents
in a hypothetical scenario, to map out potential interactions and to specify what can
be known and expressed by a science pertaining to the corresponding scenario.

of ubiquitous computing through augmented reality applications, it may be an actual physical object such as a painting in a museum or a built object, which, when seen through a cell phone screen, for example, is virtually augmented with information.

Mediation In the case a mixed object collection of real-life objects augmented with data, an S mediates the user's experience with C with the aim of helping them fulfil their information need, as opposed to 'retrieving' the object in the usual way. This clear mediation between U and C in this case is analogical to the way N is presented to O through M, i.e the mediation of a device (through visually augmenting reality) is analogical to the mediation of measurement apparatus.

Social Context A collection is formed through social processes, the laws or rules by which it is formed depend thereby on social factors. Social processes depend on pre-existing relations between social objects, objects that tend to exhibit complex internal relations, and by nature are under constant transformation. Although a C can be taken as a closed-collection, and an RS/RE situation as a closed-system, they are by nature open-systems. Given the fluid nature of social reality that contextualise the objects of C, a question arises here (Q2) as to what *kind* of objects these documents actually are. Objects in N can be particles/waves, be 'in' fields, etc., what about documents? Determining the kind of a document depends on characterising the meanings 'embedded in' it, and the meanings it takes in a social context, (Q3) does then IR's seeking to be a science depend on its ability to offer or assimilate rigorous characterisation of objects of C in context, just as particle physics has done for objects in N? While Q is taken as a closed system, and its open-system features are re-introduced by notions such as non-locality and decoherence, the open-ness in IR is more immediately apparent from O in RE or U in RS when they are observing, for example, a fast-changing subset of the internet C. IR is taken to be closed in O/T-(U-S-C) only in so far as there is a practical need to create deterministic algorithms for serving the user in a consistent tool-like way - this is IR in its purely techné sense where it has to deal with the immediate task at hand.

2.2   Types of Observation

The extent to which N of Q is available to perception is commensurate with what can be understood by interaction with it through M, thus N is ontologically further from any knowing agent in Q than C is in RE/RS, and this distance from N entails the probe-like-dynamics of experimentation. However, the U-S-C interaction is more fluid taking on a fluency akin to reading text[3] and in that sense it is more *hermeneutical* than interactional. It is the latter only from the perspective of a "dumb system" S which, like our ignorance about N, is analogically, ignorant of U (only understanding it in a simple algorithmic fashion). It

---

[3] This refers to the entirety of computer use for retrieval including browsing/selecting/saving etc, not only the actual reading of documents, further discussion with a generalisation of the notion of a retrieval application is found in [1, sec. 2.5 and ch. 3].

is perhaps even more so given the lack of laws regarding user behaviour, thus according to the way an object *appears* to agents, only S-U resembles the O-M-N. However, given that the purpose of an RE is to determine how to best serve U, in the O-(S-U/U-S-C) perspective, it is U that is the ultimate object of knowledge. Moreover, from a scientific (*episteme*) perspective (in contrast to a purely ad-hoc experimental one), it is not a particular U but properties common to several U's that concerns O (and T). Hence, if the purpose of observation is to discover such regularities and change the object of observation, then O-(S-U/U-S-C) corresponds to the O-M-N since the aim of effecting U by S and N by O is shared between IR/IS and QT respectively.

2.3   Selecting Phenomena and Making Statements

In (Stapp's interpretation of von Neumann's objective version of) the O-M-N case, the mathematics represents not the objective behaviour of elements in N but our (intersubjective) experience of this behavior, and *"[von Neumann's] objective version of QM..[considers]..both physical and phenomenal aspects..[where]..the quantum dynamical laws .. integrate the phenomenal / experiential realities into an evolving, objective, physically described universe."*[7, p192], i.e. the inter-subjective experience (and subsequent knowledge) of any O (from a T perspective) is what is to be modelled, and that model in turn is that through which N is characterized. In the QTIR case, the modelling is of that which *can* be objectively known by S in a S-U perspective. However, this integration in QT followed from an analysis of where to 'cut' the Q scene [7], a cut separates the prober from the probed, observer-side phenomena from the observed-side phenomena - (Q4) where ought one to cut in RE? That is, what extent ought we to consider the consciousness of U, its social context, instead of only its interactive behaviour in some RS? This QT inspired question corresponds in IR to questions regarding 'context' of search scenarios.

Multiple cuts Investigating the conscious experience of the user could be quite informative from a O-U perspective, although impractical from a S-U perspective. Thus multiple cuts/perspectives seem favorable, due (at least) to the dual purpose of IR and IS as techné (practical, oriented towards craft) and episteme (theoretically inclined towards phenomena classification, for example), (Q5) what would it mean to combine these multiple cuts, corresponding to different perspectives (U-S, U-S-C, S-U, etc) into a common narrative? IR/IS already deals with with the phenomena from these perspectives through user studies, matching algorithms, interaction models, usage-logs, collection models, interface designs etc. However, the question is concerned with what a synthesis would look like, and what purpose it may serve.

Analogy of Processes There are two processes, due to von Neumann, that traditionally concern Q. Process one corresponds to the probing of M/N by O (for observation) and process two to the changes in M/N following the probe [7]. Corresponding to this, there are two general categories of phenomena for QTIR to model: (1) the different types of observations and (inter)actions an agent can make and partake in, about some other agent/object (e.g. S about U or U about S/C); and (2) the changes within and between agents upon interaction, the types of possible changes, and regularities therein. For (1), the different parts of agents/objects (of U, S, C, O) require to be elaborated. Are only user interactions to be considered or 'changes in mental states' as well? In the latter case U is two parts, interaction part and cognitive part - how do these parts themselves interact with one another? The interface of S is a part, it can consist of several other parts or objects of perceptual/interactional significance, such as particular buttons or graphical layouts, but S also has an algorithmic part that interprets and reacts to interaction (this part is of particular interest to O); terms are parts of documents of some C. This aspect of modelling, corresponding to process one, is about exhaustive specification of parts and possible interactions pertaining to those parts. Modelling for (2) entails specification of the potential influence between parts (see [1, sec. 5/5.4]) and how that influence manifests, e.g. in what way can some part of U affect some part of S, or S of C, or S of U? If observing a document influences or changes a user, then in what sense is this so, is there a belief function to update, or are there other ways to indicate state change? These are process two type questions.

Types of 'scientific' inquiry Observations that can be made, in a process one context, correspond to states in which a property (part) maybe found. A researcher in O-(U-S-C) could ask whether the concept 'x' is a property/part of the information need of U and having established that it is, ask whether a set of interactions from U-S and responses in S-U, satisfy the property (and goal) of 'having-appeared' on the interface of S. It could also ask, in a more (QT-style) 'yes-no' format, whether the system's view of relevance (in S-U) takes on a particular value. A more natural question, especially pertaining to the modelling of process two, is of the form "what can S possibly know about U in the course of a RE when U exhibits a set of interactions I", or "will a particular search goal, so defined, be met..". Employing the notion of complementarity, one could ask if a particular observation-act, an act to know a property of some other agent/object prevent it from knowing some other property of that agent. That is, (Q6) how are complementary observations/interactions, between U-S for example, to be characterized - and are they interesting from an IR/IS context?

Types of scientific statements In the strict IR sense, modelling particular user related phenomena (e.g. relevant documents/terms) as it would be given to the knowledge of S, is purposed to "..furnish information that is potentially relevant to manipulation and control: they tell us how, if we were able to change the value of one or more variables, we could change the value of other variables."[8, p6], as per the manipulationist view of science. If however, the aim of IR (and IS) is to also attempt to posit relationships between objects in C and behaviours of U's, not unlike it is the task of physics to determine laws pertaining to N, then given the possibility of O-U (and O-(U-S-C/S-U)), (Q7) would not IR (in the spirit of QT) require to consider a vertically developing understanding of U? By this i mean a formal exposition of user types, interactions, and such, in addition to what is traditionally known as user-studies, and studies limited to a purely statistical-psychology based analysis; and while user-simulation is a "potentially algebraic" technique for developing an understanding of and building up a library of facts about Us, (Q8) is not its success predicated on understanding of U (and also S-U/U-S) in its full (psycho-social) context in a way analogical to the fact that Q scenes occur with a background understanding of physical concepts concerning, for example, energy and dynamics?

3   Analogy of Individual Phenomena

Why is it that physical phenomena are thought to be potentially beneficial at the psycho-social level, are they not phenomena de-contextualized by being translated between worlds? It is not that such phenomena are directly relevant, instead they serve as metaphors, aiding the conceptualisation of phenomena native to IR. They also serve a pedagogical function as we can learn to emulate in IR, the way the phenomena are dealt with by the theoretical (and experimental) apparatus in QT. And it is (at least) in these senses that the "scientific borrowing" is happening. I will briefly elaborate on superposition and entanglement in this regard.

3.1   Superposition

Superposition is a "placing above" (a 'stacking') through "moving closer" a set of objects with/to another set of objects. The ranking of objects of C as perceived on the interface of S by U in a U-S-C, is firstly a juxtaposition (a "placing near"); it is then a superposition to the extent that placing an object above another leads to one affecting the other in some way, i.e. interaction, *as it would be perceived* by a group of users, thus being an intersubjective phenomenon. To clarify, it is not documents that interact with each other but the perceptions of their contents as held by an user looking at them; i.e. documents don't have agency in RE/RS but objects in N do as they participate in causal relationships. (Q9) In what way do document perceptions affect one another when documents are juxtaposed, what is the sense of their superposition? (Q10) Can the perceptions caused by document observations be described in terms of *value of observation* ("amplitude") and *frequency of observation*? If so one can speak about two documents in a ranking or browsing path, as having the same/similar value of observation (same/similar amplitude), but one can also perhaps speak of two similar chunks of browsing - two parts that are same or close in the changes of value/amplitude, and so it could be meaningful to speak n terms of frequency.[4]

The value/amplitude function has meaning for O/U and relates to the purpose of a RS, one such function refers to the diversity approach in retrieval. If two objects are looked at in succession, and they are similar to one another, then this is a low-value reading given that the objects were retrieved as a result of the same query, and diversity is sought as the idea is to give the user a varied map

---

[4] This type of regularity may be rare in practice (see section 2.1), for a single user, but it cannot be ruled out given the multitude of ways to aggregate browsing/looking patterns for groups of users.

of meaning (see applications in [6]). In a slightly different sense, value functions can also represent whether a particular reading/browsing brings one 'closer' to some 'goal state' [1, p184-185, p200-202], thereby accommodating some types of scientific inquiry mentioned in section 2.3. There at (at least) three notions within the idea of superposition that can be imported into IR in analogical fashion (1) the undifferentiated juxtaposition of objects (as observed by U/S in U-S/S-U, or O in U-S-C and O-C), a state of potentiality as a set of possibilities (as in a wave function representation, where 'a collapse' denotes a change of belief, an actuality), (2) a juxtaposition of objects related to a perceptual value function such that observation of one can affect, in the context of the value-function, the observation of a future object, so the perceptions of objects become superposed over the course of a RS/RE; and (3) the wave-like regularity in user perceptions/behaviours where sets of behaviours/perceptions are grouped, and patterns among them (and their frequency) sought. This last notion especially pertains to patterns over time (or over objects), and at the level of modelling and expression finds itself characterized in terms state and state-changes, and possible representation by groups, see [1, ch4].

3.2   Entanglement

One notion of "entanglement" at the level of RE/RS refers to a phenomenal coupling of agents or aspects of agents, for example we could say that "the interface is entangled with the users expression.." [1, p175] in that there is "mutual adaptation of subject [U] and object [interface of S]" [3, p76]. This refers to an a priori type relationship to the extent that it can be known prior to a RE; thus, each perspective, U-S-C, U-S, O-M-N, is entangled. This entanglement is "intrinsic" to the act of "perception by mediation" (or tool-use). There is also an "a posteriori" type of "entanglement", a spontaneous coupling, such that two states/agents/objects consistently influence (depend) on each other and there is missing information as to the cause of this, more specifically "that they influence each other in a way not fully determined, determinable or deterministic." [1, p196]. This influence can happen over time, e.g. over search sessions, or over 'space', e.g. over the space of documents in C. In the latter case, it is akin to a type of latent semantics, i.e. a relationship between terms or documents which only came into view upon analysis and was otherwise hidden/latent; or in the former case, it can be used to label, for example, the phenomenon of a (peculiar) combination of user habits, such as always clicking documents with term 'x upon browsing documents containing term y even though there exist no documents in C where x and y co-occur, so it is unknown to S in RS or S/O in RE why this happens except to say "it's just the user's habit". Notice that in this second case the entanglement is by virtue of the ignorance of the observer, it concerns the epistemological in the basic picture, see section 2.

Entanglement could be used to refer to hypothetical RE situations where there are emerging relationships over the course of search scenarios (e.g. emerging relations between videos on youtube), i.e. relationships that appear at some point, and then persist for a while. It is unclear whether such a phenomena is closer in analogy to a posteriori entanglement or is better articulated as an *invariance*. I suspect that the latter is more appropriate especially when the phenomenon can be seen as invariance under transformation - i.e. user maintaining their habits in a changing environment. In general, entanglement can be used to refer to a priori relationships discovered in O-(U-S-C) or any emerging/a posteriori relations that otherwise escape sufficient explanation i.e. regularities/patterns, although whether they better resemble invariance (and emergent properties) or entanglement needs to be settled.

4   Analogies at the Level of Models and Expression

If QT's method of abstracting to mathematical structures is followed, then there is first the extraction of the basic modes of description for those QT phenomena that translate well into IR: entanglement and a priori relations, (a posteriori entanglement or) invariance, and superposition between observations. More fundamentally, these include the methods of modelling state and state change. These are more basic than phenomena as phenomena are described in terms of them. The states are those of the objects (see section 2.3), agents, and their parts. They are epistemic states, i.e. what an agent knows about some other agent at some time, and in general they all refer to the epistemic state of O or T. In the spirit of QT then, modelling begins through exposition of possibilities through toy-examples and thought-experiments (easier to do given similarity between O and U). There is at this level a mapping out of all possible (types of) states. This corresponds in RE to thinking about the possible U-S-C states, and the interactional possibilities.

Models A model of states (of a U-S-C for example) then corresponds to (mathematical) sets of objects. State change, empirical change (in belief for example), is characterised by transformations on sets. Transformations of similar types can be further abstracted as groups of transformation that work-together, i.e. the collection of transformations takes on meaning additional to the individual transformation, e.g. a particular set of user interactions indicating an 'overall high interest' in a topic where another set of transformations/interactions indicating 'overall ambiguity in interests' (from a S-U).

Expression Recall again the dual purpose of IR as a techné and episteme. Although its mainly the former, by virtue of being part of IS, it inherits the latter. This is since IS employs IR in making its statements about the world of material things and mind (and derived structures e.g. institutions such as libraries), as a scientist uses tools to make claims about the natural world.

IR can make general statements about regularities in RSs, from the perspective of an O experiencing REs , see section 2.3. It can also opt to make claims about habits of Us, the changes in C's (when C refers to the internet for example). IR also looks to classify its objects of interests: Us, Ss, Cs, U-S-Cs, and scenarios in general, into formal types, corresponding to the discursive way in which search tasks are characterised (in IS). And in this regard, it cannot do without O's knowledge of U in context, i.e. its knowledge about social reality (see section 2).

Descriptive Maths QT's notation provides a rich descriptive language, which QTIR inherits, I propose that this stands first to support the episteme purpose of IR to the extent that purpose can be realised in the (yet) closed-system paradigm of QTIR. This is so since the two processes of Q encourage an ontological and epistemological exploration (as in section 2.3), and an exploration of what kinds of statements can be made and whether regularities may develop; and so it concerns us with the subjective and intersubjective in U and O, and thereby opens up a formal door to the social, to "context". This door is one of interpretation, in the sense that the descriptive maths can be used to suggest different types of possibilities, and social reality is called upon to interpret them. Descriptive mathematics serves as a middle-language of thought and expression for IR phenomena, so that when one is using the language (playing a language game thereby) they are not too abstracted from U-S-C (and U) lest they become distant from real REs/RSs, and are often reminded to consider (algebraic) regularities across REs/RSs (i.e. episteme concerns) as opposed to focusing on 'experimental optimisation' of individual REs (e.g. techné concerns).

Empirical setup There is rich Hilbert space structure, with measures, logic, density operators and such, onto which the descriptive maths can be mapped [6]; and traditional IR is familiar with spaces and computing measures therein. However, the development of the descriptive math, and the clarifying of sense of analogy by which phenomena taken into QTIR, seem prerequisite to 'effective borrowing' at the empirical level. There are several types of measures one can explore, which appear by virtue of the empirical setup of QT being able to accommodate probabilities and distances in one framework, and due to states having a definite representation (see [1, sec 4.3.5]); but the benefits therein are only fully realized in the context of the whole project, i.e. through QTIR theories that encompass all three levels (see section 1).

5 Conclusion

QT is a scientific mirror for IR, and in comparative analysis with QT (as a result of the QTIR enterprise), certain aspects of IR are highlighted, and become part of a discursive-complex. These aspects of the discursive complex correspond particularly to general 'scientific' questions: what are the natures of phenomena in IR, how are they to be classified and enumerated, how do they relate to one another and change over the course of a search, how ought they to be characterised/measured and employed in discourse. Answers are expected to be expressed qua possibilities, in terms of mathematical structures (as per the semantic conception of science for example), but not before a careful consideration of the (inter) subjective processes they refer to. With respect to the structures and their mathematical variations, the 'culture' inherited from QT inclines the researcher (T) towards their rigorous interpretation with respect to reality (i.e. social reality), paving the way for a discourse that more accommodating of questions from a psychological and philosophical perspective, thereby potentially linking IR through QTIR to key intellectual discussions.

References

1. Arafat, S.: Foundations of Information Retrieval Inspired by Quantum Theory. PhD thesis, University of Glasgow (2008)
2. Bhaskar, R.: Feyerabend and bachelard: two philosophies of science. New Left Review 94 (1975)
3. Heelan, P.: Hermeneutical Phenomenology and the Natural Sciences. Journal of the Interdisciplinary Crossroad 1(1), 71–88 (2004)
4. Olaisen, J., Munch-Petersen, E., Wilson, P.: Information science: from the development of the discipline to social interaction. Scandinavian University Press (1995)
5. van Rijsbergen, C.J.: The Geometry Of Information Retrieval. Cambridge University Press, Cambridge (2004)
6. Song, D., Lalmas, M., van Rijsbergen, C.J., Frommholz, I., Piwowarski, B., Wang, J., Zhang, P., Zuccon, G., Bruza, P.D., Arafat, S., et al.: How quantum theory is developing the field of Information Retrieval. In: AAAI Fall Symposium Series (2010)
7. Stapp, H.: Mind in the Quantum Universe. In: Tymieniecka, A.-T., Grandpierre, A. (eds.) Analecta Husserliana CVII, pp. 189–198
8. Woodward, J.: Making things happen: A theory of causal explanation. Oxford University Press, USA (2003)

A Hierarchical Sorting Oracle

Luís Tarrataca* and Andreas Wichert

GAIPS/INESC-ID
Department of Computer Science, Instituto Superior Técnico
{luis.tarrataca, andreas.wichert}@ist.utl.pt

Abstract. Classical tree search algorithms mimic the problem solving capabilities traditionally performed by humans. In this work we propose a unitary operator, based on the principles of reversible computation, focusing on hierarchical tree search concepts for sorting purposes. These concepts are then extended in order to build a quantum oracle which, combined with Grover's quantum algorithm, can be employed as a quantum hierarchical search mechanism whilst taking advantage of a quadratic speedup. Finally, we show how the developed model can be extended in order to perform a N-level depth-limited search.

Keywords: quantum search; tree search; artificial intelligence.

1 Introduction

Tree search algorithms assume a crucial role in artificial intelligence where they are employed to model problem solving behaviour. Typically, such problems can be described by a tuple (S_i, S_g, R) where S_i represents a finite set of initial states, R a finite set of actions and S_g a finite set of goal states. The objective of such algorithms consists in determining a sequence of actions leading from an initial state to a goal state. A wide range of problems has been formulated in terms of hierarchical search procedures *e.g.* game playing programs and robot control systems. Such behaviour requires the ability to determine what state is obtained after applying an action to a given state. This process is illustrated in Figure 1 where a set of possible actions, respectively $R = \{0, 1\}$, is applied to a root node A producing in the process a binary tree. The cardinality of the set of available actions is also referred to as the branching factor b. At a search depth level d there exist a total of b^d leaf nodes. Each leaf node translates into the state reached after having applied d actions, *e.g.* node I is reach after applying actions 0, 0 and 1. We will refer to set of actions leading to a leaf node as the path taken during the tree search.

Grover's quantum search algorithm [1] allows for a quadratic speedup to be obtained in search procedures. The algorithm performs a generic search for n-bit

---

* This work was supported by Fundação para a Ciência e Tecnologia (FCT) (INESC-ID multiannual funding) through the PIDDAC Program funds and FCT grant DFRH - SFRH/BD/61846/2009.

D. Song et al. (Eds.): QI 2011, LNCS 7052, pp. 172–181, 2011.

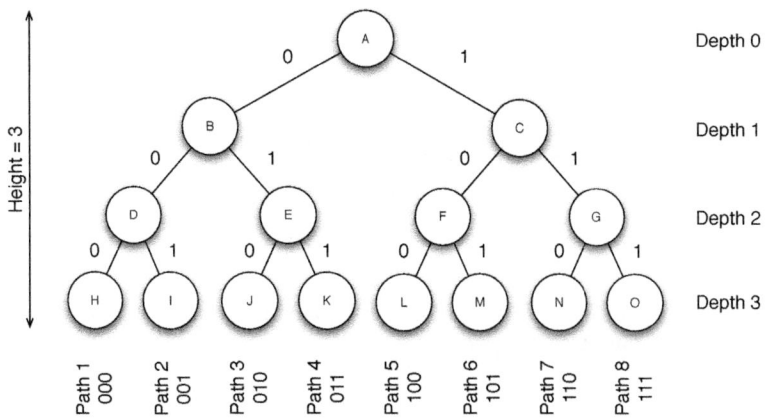

Fig. 1. The possible paths for a binary search tree of depth 3

solutions amongst the 2^n possible combinations by employing the quantum superposition principle alongside an oracle O in order to query many elements of the search space simultaneously. The oracle is responsible for determining which strings correspond to solutions and it should be able to do so in polynomial time. This behaviour is similar to the NP class of problems whose solutions are verifiable in polynomial time $O(n^k)$ for some constant k, where n is the size of the input's problem. Oracle O behaviour can be formulated as presented in Expression 1, where $|x\rangle$ is a n-bit query register, $|c\rangle$ is a single bit answer register where the output of $g(x)$ is stored. Function $g(x)$ is responsible for checking if x is a solution to the problem, outputting value 1 if so and 0 otherwise. Grover's original idea only focused on developing a generic search mechanism and did not have hierarchical search in mind. In this work we consider the impact of incorporating classical search concepts alongside Grover's algorithm into a hybrid quantum search system capable of solving instances of the hierarchical sorting problem.

$$O : |x\rangle|c\rangle \mapsto |x\rangle|c \oplus g(x)\rangle \tag{1}$$

The remainder of this work is organized as follows: Section 2 introduces the concepts of the hierarchical sorting problem; Section 3 presents the required reversible circuitry for our proposition alongside an oracle mapping capable of being integrated with Grover's algorithm; Section 4 discusses how such an oracle can be applied alongside Grover's algorithm and how our proposition differs from quantum random walks on graphs; Section 5 presents the conclusions of our work.

2   Sorting

The sorting problem may be defined in terms of the application of a problem-specific set of actions with the objective of determining a sequence of actions that

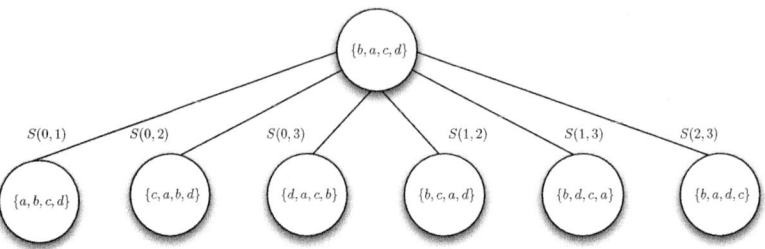

Fig. 2. A search of depth 1 with a branching factor $b = |R| = 6$ applied to an initial state $\{b, a, c, d\}$ and goal state $\{a, b, c, d\}$

produces a goal state. For some problems, the action set may convey an increasing element order, whilst for others the final arrangement may only be expressed through condition-action pairs. For some problems the only viable procedure consists in performing an exhaustive examination of all possible actions until goal states are found. *E.g.* suppose we wish to sort a list containing elements of an alphabet $\sum = \{a, b, c, d\}$ and that the dimension of the list, E, is fixed to four elements. In each computational step we can perform operation $S(x, y)$, responsible for switching the elements in position x and y. If repetitions are not allowed then is is possible to check that a total of $\binom{|\sum|}{2} = \binom{4}{2}$ possible combinations exist, where $|\sum|$ represents alphabet length. Accordingly, we are able to define the set of possible actions $R = \{S(0, 1), S(0, 2), S(0, 3), S(1, 2), S(1, 3), S(2, 3)\}$, and apply it an initial state, as illustrated in Figure 2.

3   Oracle Development

Changes occurring to a quantum state can be described with quantum circuits containing wires and elementary quantum gates to carry around and manipulate quantum information [2]. Mathematically, state evolution can be expressed unitary operators. A matrix A is said to be unitary if A's transpose complex conjugate, denoted by A^{*^T}, or simply by $A^\dagger$, is also the inverse matrix of A [3]. In this notation each matrix column describes the transformation suffered at a specific column index, *i.e.* a permutation. These concepts are related to reversible computation theory, ergo our approach relies on developing a reversible circuit capable of sorting the 4-length list element presented in Section 1. Therefore, we need to represent the overall state in a binary fashion. More specifically, $\lceil \log_2 |\sum| \rceil = \lceil \log_2 4 \rceil = 2$ bits are required to encode the symbols of the alphabet, each of which can be represented as presented in Table 1. This implies that a total of 8 bits will be employed to represent each list. Let Table 2 represent the encodings for the root state and the goal state associated with the sorting example of Figure 2. Conceptually, our reversible circuit will require the ability to: (1) determine if a state is a goal state; and (2) given a state and an action determine the new state obtained. These two requirements will be discussed,

Table 1. Binary encoding for each symbol of $\sum$

b_0 b_1	Element
0 0	a
0 1	b
1 0	c
1 1	d

Table 2. Binary encodings for the initial and goal states of Figure 2

Position	0		1		2		3	
Bits	b_0	b_1	b_2	b_3	b_4	b_5	b_6	b_7
$\{b, a, c, d\}$	0	1	0	0	1	0	1	1
$\{a, b, c, d\}$	0	0	0	1	1	0	1	1

respectively, in Section 3.1 and Section 3.2. Section 3.3 presents the details of the overall circuit.

3.1 First Requirement

Tackling the first requirement requires developing a gate capable of receiving as an argument a binary string representing the state and testing if it corresponds to a goal state. This computational behaviour can be represented through an irreversible function f, as illustrated in Expression 2. It is possible to obtain a reversible mapping of an irreversible function f with the form presented in Expression 3, where x represents the input and c an auxiliary control bit [4].

$$f(\underbrace{b_0, b_1, b_2, b_3, b_4, b_5, b_6, b_7}_{\text{state}}) = \begin{cases} 1 \text{ if state } \in S_g \\ 0 \text{ otherwise.} \end{cases} \tag{2}$$

$$(x, c) \mapsto (x, c \oplus f(x)) \tag{3}$$

From Expression 3 we know that the inputs should also be part of the outputs. The only issue is due to the result bit, which requires that a single control bit be provided as an input. Therefore, any potential gate would require 9 input and output bits, 8 of which are required for representing the state and 1 bit serving as control. This gate, which we will label as the goal state unitary operator, is illustrated in Figure 3. Table 3 showcases the gate's behaviour for a selected number of states, where $f(b)$ denotes $f(b_0, b_1, b_2, b_3, b_4, b_5, b_6, b_7)$. Notice that when the gate determines that the input state $\in S_g$ it effectively switches the control bit, as highlighted in Table 3. Mathematically, we need to specify the set of column permutations. Let T denote the unitary operator responsible for implementing the behaviour of function f. T is a matrix with dimensions $2^9 \times 2^9$. From Table 3 it should be clear that only two input states map onto other states rather than themselves. Namely, $T|54\rangle \rightarrow |55\rangle$ and $T|55\rangle \rightarrow |54\rangle$. Accordingly, the 54th column of T should permute to state $|55\rangle$, and the 55th

Table 3. A selected number of results from the truth table of the goal state unitary operator

Inputs									Outputs								
b_0	b_1	b_2	b_3	b_4	b_5	b_6	b_7	c	b_0	b_1	b_2	b_3	b_4	b_5	b_6	b_7	$c \oplus f(b)$
0	0	0	1	1	0	1	1	0	0	0	0	1	1	0	1	1	1
0	0	0	1	1	0	1	1	1	0	0	0	1	1	0	1	1	0
:	:	:	:	:	:	:	:	:	:	:	:	:	:	:	:	:	:
0	1	1	0	1	1	0	0	0	0	1	1	0	1	1	0	0	0
0	1	1	0	1	1	0	0	1	0	1	1	0	1	1	0	0	1
:	:	:	:	:	:	:	:	:	:	:	:	:	:	:	:	:	:
1	1	1	0	0	1	0	0	0	1	1	1	0	0	1	0	0	0
1	1	1	0	0	1	0	0	1	1	1	1	0	0	1	0	0	1

column map to state $|54\rangle$. All other remaining states would continue to map onto themselves.

3.2   Second Requirement

The second requirement combined alongside with Expression 3 implies that the new state should be presented alongside the original one. Additionally, we are interested in applying a switch action if and only if the input state $\notin S_g$. As a consequence, we can opt to develop a new function g which includes in its definition a reference to function f. Our main concern resides in how to output the new state in a reversible manner since we are interested in having 8 result bits representing the new state. Expression 3 can be extended in order to accommodate any number of control bits, as illustrated by Expression 4 where c_i are control bits, and $f(x) = (y_0, y_1, \cdots, y_{n-1})$ with $y_i \in \{0, 1\}$. Function g is responsible for producing the new state by taking into account the current state and four bits, respectively (m_0, m_1) and (m_2, m_3), representing, respectively, the arguments x and y of the switching function $S(x, y)$. Accordingly, let $g : \{0, 1\}^{12} \to \{0, 1\}^8$ with $g(b, m) = (y_0, y_1, y_2, y_3, y_4, y_5, y_6, y_7)$, where b denotes the input state $(b_0, b_1, b_2, b_3, b_4, b_5, b_6, b_7)$, m the positional bits (m_0, m_1, m_2, m_3) and $(y_0, y_1, y_2, y_3, y_4, y_5, y_6, y_7)$ the resulting state. Then, g's behaviour has the form presented in Expression 5. The corresponding gate therefore has (1) 8 input and output bits for the current state; (2) 4 input and output bits describing the switch positions; and (3) 8 control and result bits in order to account for the new state. The reversible gate, which we will refer to as the switch element operator M, is depicted in Figure 3. The corresponding unitary operator M is a matrix of dimension $2^{8+4+8} \times 2^{8+4+8}$ which can be built in a similar way to T.

$$(x, c_0, c_1, \cdots, c_{n-1}) \mapsto (x, c_0 \oplus y_0, c_1 \oplus y_1, \cdots, c_{n-1} \oplus y_{n-1}) \tag{4}$$

$$g(b, m) = \begin{cases} (b_2, b_3, b_0, b_1, b_4, b_5, b_6, b_7) & \text{if } f(b) = 0 \text{ and } m = (0,0,0,1) \\ (b_4, b_5, b_2, b_3, b_0, b_1, b_6, b_7) & \text{if } f(b) = 0 \text{ and } m = (0,0,1,0) \\ (b_6, b_7, b_2, b_3, b_4, b_5, b_0, b_1) & \text{if } f(b) = 0 \text{ and } m = (0,0,1,1) \\ (b_0, b_1, b_4, b_5, b_2, b_3, b_6, b_7) & \text{if } f(b) = 0 \text{ and } m = (0,1,1,0) \\ (b_0, b_1, b_6, b_7, b_4, b_5, b_2, b_3) & \text{if } f(b) = 0 \text{ and } m = (0,1,1,1) \\ (b_0, b_1, b_2, b_3, b_6, b_7, b_4, b_5) & \text{if } f(b) = 0 \text{ and } m = (1,0,1,1) \\ (b_0, b_1, b_2, b_3, b_4, b_5, b_6, b_7) & \text{otherwise} \end{cases} \quad (5)$$

3.3   General Circuit

By combining both the switch elements and the goal state gates we are now able to verify if a goal state has been reached after switching two elements. The switch elements operator M already incorporates in its design a test for determining if the gate should be applied or not. Accordingly, we only need to check if the final state obtained corresponds to a goal state. This process is illustrated in Figure 3 where a switch operator M is employed alongside a goal state operator T, where res has the value presented in Expression 6.

$$res = c_8 \oplus f(c_0 \oplus y_0, c_1 \oplus y_1, c_2 \oplus y_2, c_3 \oplus y_3, c_4 \oplus y_4, c_5 \oplus y_5, c_6 \oplus y_6, c_7 \oplus y_7) \quad (6)$$

Algebraically, the overall circuit behaviour can be expressed as presented in Expression 7, where $I^{\otimes(8+4)} = I \otimes I \otimes \cdots \otimes I$ repeated 12 times, since operator T should only take into consideration bits $c_0, c_1, \cdots, c_8$. The associated unitary operator, respectively presented in Expression 7, acts on Hilbert space $\mathcal{H} = H_b \otimes H_m \otimes H_c$, where H_b is the Hilbert space spanned by the basis states employed to encode the state configuration bits $b = b_0, b_1, \cdots, b_7$, H_m is the Hilbert space spanned by the basis states employed to represent the set of permutations, and H_c is the Hilbert space spanned by the auxiliary control bits.

$$(I^{\otimes 12} \otimes T)M|b_0, b_1, \cdots, b_7, m_0, m_1, m_2, m_3, c_0, c_1, \cdots, c_8\rangle \quad (7)$$

This strategy can be extended in order to apply any number of switch operators, where the output of a switch gate is provided as input to another switch operator. In doing so, we add a guarantee that, if possible, another element permutation is applied to the input state. More specifically, in order to represent each element of the alphabet we require $e = \lceil \log_2 |\sum| \rceil$ bits. Let E represent the element list to be sorted, then an adequate encoding for E will require $b = |E| \times e$ bits, where $|E|$ denotes the list size. Additionally, specifying a list position involves $p = \lceil \log_2 |E| \rceil$ bits. Each switch operator M will thus require a total of $b + p + p + b = 2(b + p)$ input and output bits, and each goal state gate T will require a total of $b + 1$ input and output bits. How many bits will be required by the circuit? Suppose we wish to apply m permutation, i.e. apply operator M a total of m times. The first operator M_1 requires $2(b + p)$ bits. Since a part of M_1 outputs will be provided as input to M_2 an additional $b + 2p$ bits will be added to

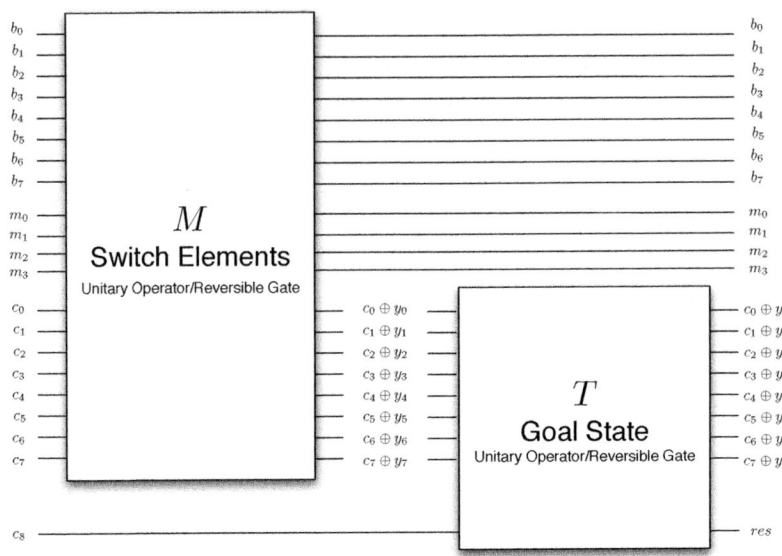

Fig. 3. The reversible circuit responsible for performing the depth-limited search of Figure 2

the circuit. If we extend this reasoning to m applications of M then it is possible to conclude that $2(b+p) + (m-1)(b+2p)$ bits will be required to perform the switching operations. Since operator T requires a single control bit this implies that the overall circuit employs a total of $n = 2(b+p) + (m-1)(b+2p) + 1$ bits.

Of these n bits $c = n - (b + m \times 2p) = mb + 1$ bits are control, or auxiliary, bits. Furthermore, the sequence of bit indexes after which a switch operator M should be applied is $V = \{0, b+2p, 2(b+2p), 3(b+2p), \cdots, (m-1)(b+2p)\}$. Based on these statements we can describe a general formulation for a sorting circuit C employing operators M and T, as illustrated in Expression 8. Unitary operator C would act on an input register $|x\rangle$ conveying information regarding the initial state, the set of permutations and also the auxiliary control bits. Accordingly, operator C would act upon a Hilbert space $\mathcal{H}$ spanned by the computational basis states required to encode x. Notice that this approach is equivalent to performing a depth-limited search, one whose number of switch operators T would grow linearly with the depth.

$$C = (I^{\otimes m(2p+b)} \otimes T) \prod_{k \in V} (I^{\otimes k} \otimes M) \tag{8}$$

Expression 8 needs to be further refined in order to be in conformity with the oracle formulation of Expression 1. which effectively means that all the original inputs, excluding bit c, should also be part of the outputs. This means that the circuit presented in Figure 3 should somehow undo their computation and

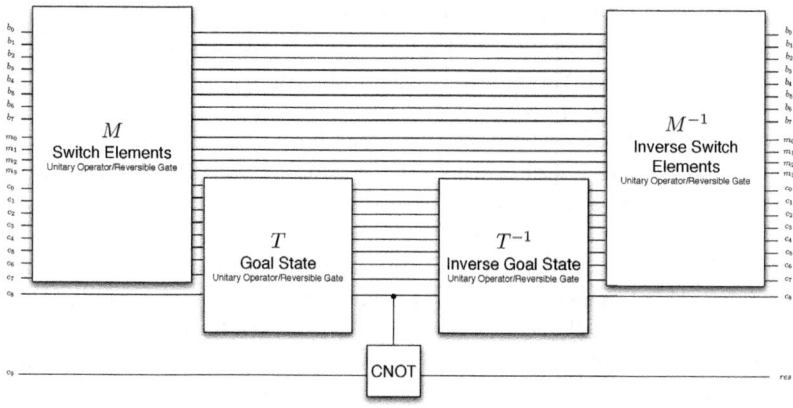

Fig. 4. The oracle formulation of the depth-limited search circuit of Figure 3

then store the overall conclusion in an output register, an operation which can be performed by employing a CNOT gate. This behaviour can be obtained by building a mirror circuit, C^{-1}, where each component is the inverse operation of original circuit. Then, with both circuits developed, it is just a matter of establishing the appropriate connections, *i.e.* the outputs of the original circuit are provided as inputs to the mirror. The application of these requirements to the reversible circuit of Figure 3 is presented in Figure 4. The circuits output is presented in Expression 9 where *res* has the value shown in Expression 6. If the input register $|b\rangle|m\rangle|c\rangle$ is relabeled as $|x\rangle$ then Expression 9 is equivalent to Expression 1.

$$O : \underbrace{|b\rangle|m\rangle|c\rangle}_{\text{input}} \quad \underbrace{|c_9\rangle}_{\text{oracle's control bit}} \quad \mapsto |b\rangle|m\rangle|c\rangle|c_9 \oplus res\rangle \tag{9}$$

Alternatively, we can state this result in more general terms by employing unitary operator C, presented in Expression 8, as showcased by Expression 10. In both cases the Hilbert space $\mathcal{H}$ of the input register is augmented with the basis states required to encode the additional auxiliary control bit, accordingly $\mathcal{H} = \mathcal{H}_b \otimes \mathcal{H}_m \otimes \mathcal{H}_c \otimes \mathcal{H}_{c_{mb+2}}$.

$$O = C^{-1}(I^{\otimes 2(b+p)+(m-1)(b+2p)}CNOT)C|b\rangle|m\rangle|c\rangle|c_{mb+2}\rangle \tag{10}$$

4   Final Considerations

Overall, our reversible circuit and the associated oracle O can be perceived as employing a binary string of the form $|b_1b_2b_3b_4b_5b_6b_7b_8 \quad r_1r_2\cdots r_N\rangle$, where r_i represent a sequence of permutations. Accordingly, we are now able to employ Grover's algorithm alongside oracle O and a superposition $|\psi\rangle$. The exact form of $|\psi\rangle$ depends on the specific task at hand, *e.g.* (1) we may be interested in only building a superposition of all possible permutations, a behaviour similar to the depth-limited search presented in Figure 2, or (2) we may set $|\psi\rangle = H^{\otimes k}|0\rangle^{\otimes k}$, where k is the number of bits employed by the input state $|b\rangle|m\rangle|c\rangle$, effectively allowing us to search all possible combinations of initial states and permutations simultaneously. After Grover's algorithm has been applied and upon measuring the superposition state we obtain a state containing the sequence of permutations leading up to a goal state. From a tree search perspective this process can be viewed as a depth-limited search. Classical search strategies require $O(b^d)$ time, where b is the branching factor and d the depth of a solution. If we only take into consideration the dimension of the search space then such a quantum hierarchical search strategy would allow this time to be reduced to $O(\sqrt{b^d})$, effectively cutting the depth factor in half. However, this is a best case scenario since it assumes that the bit encoding strategy always produces viable paths, which is not always true depending on the dimension of the search space or when non-constant branching factors are employed (please refer to [5] for more details).

Finally, from a graph perspective, it is possible to establish some links between the concepts discussed and quantum random walks on graphs. Quantum random walks are the quantum equivalents of their classical counterparts ([6] provides an excellent introduction to the area). Quantum random walks were initially approached in [7], [8], [9] and in one-dimensional terms, *i.e.* walk on a line. These concepts were then extended to quantum random walks on graphs in [10], [11], and [12]. Quantum random walks can also provide a probabilistic speedup relatively to their classical parts, namely the hitting time for some specific graphs, *i.e.* the time it takes to reach a certain vertex B starting from a vertex A, can be shown to be exponentially smaller [13]. However, these approaches only focus on graph transversal through a simultaneous selection of all possible edges at any given node, a procedure which is applied through the superposition principle. In contrast, our approach focuses on a simultaneous evaluation of all possible path up to a depth level d with a focus on (1) finding states $\in S_g$ and (2) determining the path leading up to these states.

5   Conclusions

In this work we presented a possible model for a depth-limited search with an emphasis on sorting. The proposed model can be viewed as an hybrid between a pure quantum search mechanism, such as the one detailed in Grover's algorithm, and a classical search system. By combining these concepts we are able to hierarchically search through all possible combinations quadratically faster than its classical counterparts. Our proposal placed a strong emphasis on determining the set of actions leading up to a target node, since this a crucial task for many artificial intelligence applications. Our approach can be also perceived as performing hierarchical search by exploiting the NP class of problems.

References

1. Grover, L.K.: A fast quantum mechanical algorithm for database search. In: STOC 1996: Proceedings of the Twenty-Eighth Annual ACM Symposium on Theory of Computing, pp. 212–219. ACM, New York (1996)
2. Deutsch, D.: Quantum computational networks. Proceedings of the Royal Society of London A 425, 73–90 (1989)
3. Hirvensalo, M.: Quantum Computing. Springer, Heidelberg (2004)
4. Kaye, P.R., Laflamme, R., Mosca, M.: An Introduction to Quantum Computing. Oxford University Press, USA (2007)
5. Tarrataca, L., Wichert, A.: Tree search and quantum computation. Quantum Information Processing, 1–26 (2010), doi:10.1007/s11128-010-0212-z
6. Hughes, B.D.: Random Walks and Random Environments. Random Walks, vol. 1. Oxford University Press, USA (1995)
7. Aharonov, Y., Davidovich, L., Zagury, N.: Quantum random walks. Phys. Rev. A 48(2), 1687–1690 (1993)
8. Meyer, D.: From quantum cellular automata to quantum lattice gases. Journal of Statistical Physics 85(5), 551–574 (1996)
9. Nayak, A., Vishwanath, A.: Quantum walk on the line. Technical report, DIMACS Technical Report (2000)
10. Farhi, E., Gutmann, S.: Quantum computation and decision trees. Phys. Rev. A 58(2), 915–928 (1998)
11. Hogg, T.: A framework for structured quantum search. Physica D 120, 102 (1998)
12. Aharonov, D., Ambainis, A., Kempe, J., Vazirani, U.: Quantum walks on graphs. In: Proceedings of ACM Symposium on Theory of Computation (STOC 2001), pp. 50–59 (July 2001)
13. Childs, A.M., Cleve, R., Deotto, E., Farhi, E., Gutmann, S., Spielman, D.: Exponential algoritmic speedup by quantum walk. In: Proceedings of the 35th ACM Symposium on Theory of Computing (STOC 2003), pp. 59–68 (September 2003)

Quantum-Like Paradigm: From Molecular Biology to Cognitive Psychology

Masanari Asano[1], Masanori Ohya[1], Yoshiharu Tanaka[1], Ichiro Yamato[2],
Irina Basieva[3,*], and Andrei Khrennikov[3]

[1] Department of Information Sciences, Tokyo University of Science
Yamasaki 2641, Noda-shi, Chiba, 278-8510 Japan
[2] Department of Biological Science and Technology
Tokyo University of Science
Yamasaki 2641, Noda-shi, Chiba, 278-8510 Japan
[3] International Center for Mathematical Modelling
in Physics and Cognitive Sciences
Linnaeus University, Växjö, S-35195 Sweden

Abstract. We present the quantum-like paradigm for biology, cognitive psychology, and modeling of brain's functioning. By this paradigm contextuality of biological processes induces violation of laws of classical (Kolmogorovian) probability, starting with the fundamental law of total probability. New nonclassical models have to be used for mathematical modeling of contextual phenomena.

Keywords: Quantum probability, contextuality, formula of total probability, quantum-like models, biology, cognitive science, psychology.

1 Introduction

The idea that the mathematical formalism of quantum information theory can be used to describe information processes in the brain was elaborated in a series of papers, see [1]–[16]. This approach is based on fundamental conjecture that the real physical brain developed an ability to represent the probabilistic information in complex linear space, by complex vectors (pure quantum-like mental states) or more generally density operators (mixtures of pure quantum-like mental states).

A few years ago J. Busemeyer et al. [5] noticed that quantum-like models of decision making can be used to explain disjunction effect in experiments of cognitive psychology, see also Khrennikov [9]. In particular, he reproduced statistical data from experiments of the Prisoner's Dilemma (PD) type. Moreover, it was shown [5], [14], [8] that it is difficult to construct a classical Markovian model reproducing the aforementioned experimental data. At the same time the authors of [8] constructed a *quantum Markov chain* reproducing statistical data from cognitive psychology. In this paper we discuss a quantum-like model of decision making (and more generally processing of information by the brain)

---

* Postdoc fellowship of Swedish Institute.

D. Song et al. (Eds.): QI 2011, LNCS 7052, pp. 182–191, 2011.

which is based on decoherence dynamics, see section 4. One of the most important nonclassical features of quantum-like models is *interference of probabilities.* Interference can be either constructive or destructive. In the latter case (which will be considered in this paper) the reaction of a system to one factor, say B_+, can destroy its reaction to another factor, say B_-. Thus the presence of both factors, $B = B_+ \cup B_-$, can, in principle, minimize practically to zero the activity induced by B_+. Such destructive interference is well known in quantum physics.

As was mentioned, interference effects can be demonstrated by cognitive systems: e.g., in experiments with recognition of ambiguous figures [10] and in experiments on disjunction effect. In this paper we shall show that quantum-like interference (at least destructive) can be found in even in molecular biology; in particular, as effects of activity of genetic systems. A possibility that not only humans, but even animals can "behave in quantum-like way" was discussed in [4]. However, it was always emphasized that quantum-like behaviour is a feature of advanced cognitive systems having the nervous system of high complexity. In this note we consider the simplest biological system, a cell, and we shall see that it can exhibit (under some special contexts) quantum-like behaviour.

One of complications in the application of quantum-like probabilistic models outside of physics is that the standard calculus of quantum probabilities which is applicable to e.g. photons and electrons is too restrictive to describe probabilistic behaviour of biological systems. Biological systems are not only nonclassical (from the probabilistic viewpoint), but they are even "worse" than quantum systems. They react on combinations of incompatible contexts by exhibiting stronger interference than quantum physical systems. Instead of standard trigonometric interference of the $\cos\theta$-type which is well known in quantum physics, *hyperbolic interfernce* of the $\cosh\theta$-type can be exhibited in experiments with cognitive systems. Experiments of the later type cannot be described by the standard mathematical formalism of QM. A generalization of the QM-formalism based on so called hyperbolic amplitudes should be applied [15].

In the experiment discussed in this paper gene expression generates hyperbolic interfernce, i.e., interference which is essentially stronger than the standard quantum-like interference. In any event the data collected in this experiment [17] violates basic laws of classical probability theory.

2   Classical Law of Total Probability and Its Quantum-Like Modification

Consider two disjoint events, say B_+ and B_-, such that $P(B_+ \cup B_-) = 1$ (the probability of realization of either B_+ or B_- equals to 1) and consider any event A. Then one of basic laws of classical probability can be expressed in the form of the formula of total probability

$$P(A|B_+ \cup B_-) = P(B_+)P(A|B_+) + P(B_-)P(A|B_-), \tag{1}$$

where the conditional probability of one event with respect to another is given by the Bayes formula:

$$P(A|H) = P(A \cap H)/P(H), \tag{2}$$

for H with $P(H) > 0$. We do not discuss here applications of these rules in Bayesian analysis of statistical data; they are well known.

For statistical data obtained in experiments with quantum systems, this formula is violated, see [15] for a popular exposition. Instead of the classical formula of total probability (1), QM uses its perturbed version ("the formula of total probability with an interference term"):

$$P(A|B_+ \cup B_-) = P(B_+)P(A|B_+) + P(B_-)P(A|B_-)$$
$$+ 2\cos\theta\sqrt{P(B_+)P(A|B_+)P(B_-)P(A|B_-)}, \tag{3}$$

where θ is a phase vector. In physics this angle has a natural geometric interpretation. However, already in cognitive science the geometric interpretation of phase is impossible (or at least unknown). In [3] it was proposed to interpret the phase as a measure of incompatibility of events. Mathematically incompatibility is described as impossibility to use Boolean algebra for these events or in other words set-theoretical representation.

Already in quantum physics the event interpretation of $B_\pm$ in the formula of total probability is misleading. In real experiments, these are not events, but various experimental contexts. In applications to biology it is also useful to proceed with contextual terminology, especially in experimental situations which are characterized by violation of Bayes formula (2). Therefore we prefer to call probabilities $P(A|B_\pm)$ not conditional, but contextual.

The constructive wave function approach [15], [9] provides a possibility to reconstruct the wave function (in experiments with quantum systems), the complex probabilistic amplitude. We have, see [15],

$$\psi_A = \alpha + e^{i\theta}\beta, \tag{4}$$

where

$$\alpha = \sqrt{P(B_+)P(A|B_+)}, \beta = \sqrt{P(B_-)P(A|B_-)} \tag{5}$$

and the phase θ can be found from the "coefficient of interference"

$$\lambda_A = \frac{P(A|B_+ \cup B_-) - (P(B_+)P(A|B_+) + P(B_-)P(A|B_-))}{2\sqrt{P(B_+)P(A|B_+)P(B_-)P(A|B_-)}}. \tag{6}$$

We remark that, for quantum physical systems, the magnitudes of coefficients of interference are always bounded by 1,

$$|\lambda_A| \leq 1. \tag{7}$$

For statistical data, collected in quantum physical experiments, the phase is given by

$$\theta = \arccos \lambda_A. \tag{8}$$

We state again that the coefficient of interference λ_A can be found on the basis of experimental data (this is the essence of the constructive wave function approach [15]). The nominator of (6) gives a measure of nonclassicality of data: this is the magnitude of violation of the law of total probability; the denominator is simply a normalization coefficient.

In the absence of the experimental data the ψ-function can be obtained e.g. from the evolution equation, Schrödinger's equation. If the complex probabilistic amplitude is known then probability can be calculated with the aid of the basic formula of quantum probability, Born's rule:

$$P(A|B_+ \cup B_-) = |\psi_A|^2 = |\alpha + e^{i\theta}\beta|^2. \tag{9}$$

If $\theta \neq \pi/2$, then $P(A|B_+ \cup B_-) \neq |\alpha|^2 + |\beta|^2$. The presence of the phase θ induces interference

$$P(A|B_+ \cup B_-) = |\alpha|^2 + |\beta|^2 + 2\cos\theta|\alpha||\beta|.$$

The same approach can be used not only for quantum physical systems, but for biological systems demonstrating nonclassical probabilistic behavior, see [15], [9] for applications to cognitive systems. Instead of probabilities, one operates with wave functions, probabilistic amplitudes.

As was mentioned in introduction, biological systems can demonstrate even stronger violation of the formula of total probability than quantum physical systems, i.e., the coefficient of interference λ_A, see (6), can be larger than 1. In such situations the modified formula of total probability has the form

$$P(A|B_+ \cup B_-) = P(B_+)P(A|B_+) + P(B_-)P(A|B_-)$$

$$\pm 2\cosh\theta\sqrt{P(B_+)P(A|B_+)P(B_-)P(A|B_-)}, \tag{10}$$

i.e., the hyperbolic cosine has to be used. This type of interference was found for cognitive systems [9].

The constructive wave function approach can be generalized to the hyperbolic case. Let us consider the algebra of hyperbolic numbers: $z = x + jy$, where x, y are real numbers and the imaginary element j is such that $j^2 = 1$. Then the formula of total probability with the hyperbolic interference term, see (10), induces the representation of the probability by the hyperbolic amplitude:

$$\psi_A = \alpha \pm e^{j\theta}\beta, \tag{11}$$

where the coefficients are again given by (5), θ is a "hyperbolic phase". The latter can be found (similar to the usual "trigonometric phase"), see (8), as

$$\theta = \arccos|\lambda_A|. \tag{12}$$

The sign in (11) is determined by the sign of the coefficient of interference λ_A.

Generalization of Born's rule (13) gives the representation of the probability as the squared amplitude:

$$P(A|B_+ \cup B_-) = |\psi_A|^2 = |\alpha \pm e^{j\theta}\beta|^2 = |\alpha|^2 + |\beta|^2 \pm 2\cosh\theta|\alpha||\beta|. \tag{13}$$

The application of this general framework to microbiology is not totally straightforward. Sometimes it is difficult to determine probabilities $P(B_\pm)$ in experiments with cells. Therefore the direct test of the formula of total probability (1) is not possible (or it requires additional experiments). However, this is not a problem, because the formula (1) is a consequence of a more fundamental law of classical theory of probability, namely, the law of additivity of probabilities. We recall the derivation of (1). It can be found on the first pages of any textbooks on probability theory:

$$P(A|B_+ \cup B_-) = P(A \cap (B_+ \cup B_-)) = P(A \cap B_+) + P(A \cap B_-), \qquad (14)$$

which is a consequence of additivity of probability. This is the basic law. Then, to obtain (1), one does the formal algebraic transformation to conditional probabilities:

$$P(A|B_+ \cup B_-) = P(B_+)\frac{P(A \cap B_+)}{P(B_+)} + P(B_-)\frac{P(A \cap B_-)}{P(B_-)}.$$

Therefore it is reasonable to test the basic law of additivity of classical probability (14) whose violation implies violation of the formula of total probability which is used in Bayesian analysis of statistical data. We now can easily rewrite all above formulas on complex and more general probabilistic amplitudes by placing

$$P(B_\pm)P(A|B_\pm) \mapsto P(A \cap B_\pm). \qquad (15)$$

We point out that in experimental studies typically A is determined by values of a random variable, say ξ, which are measured in the experiment. In the simplest case ξ is dichotomous, e.g., $\xi = \pm 1$, and A can be chosen either as $A_+ = \{\xi = +1\}$ or as $A_- = \{\xi = -1\}$.

3   Violation of the Law of Total Probability in Microbiology: Glucose Effect on E. Coli Growth

Our considerations are based on an article reporting the glucose effect on *E. coli* (*Escherichia coli*) growth, see [17]. There was measured the β-galactosidase activity at certain growth phase: grown in the presence of 0.4% lactose, 0.4% glucose, or 0.4% lactose + 0.1% glucose. The activity is represented in Miller units (enzyme activity measurement condition). There was obtained the probabilistic data: 0.4% glucose, 33 units; 0.4% lactose, 2920 units; 0.4% lactose +0.1% glucose, 43 units.

We recall that by full induction, the activity reaches to 3000 units. We want to represent these data in the form of contextual probabilities and put them into the formula of total probability.

We introduce a random variable, say ξ, which describes the level of activation. We also consider two contexts: L – the presence of molecules of lactose and

G – the presence of molecules of glucose. The experimental data provide the contextual (conditional) probabilities

$$P(\xi = +1|L) = \frac{2920}{3000} \approx 0.973, \; P(\xi = +1|G) = \frac{33}{3000} \approx 0.011.$$

Consider now the context $L \cup G$ of the presence of molecules of lactose and glucose. In classical probability theory the set-theoretical description is in usage. We can represent L as the set of lactose molecules and G as the set of glucose molecules and, finally, C as (disjoint) union of these sets. (Of course, there are other types of molecules. However, we ignore them, since the random variable ξ depends only of the presence of lactose and glucose.)

We have

$$P(\xi = +1|L \cup G) = \frac{43}{3000} \approx 0.014$$

In the classical probabilistic framework we should obtain the equality (14), a consequence of the law of additivity of probabilities:

$$P(\xi = +1|L \cup G) = P(\xi = +1 \cap L) + P(\xi = +1 \cap G). \tag{16}$$

By puting the data into (16) we obtain

$$0.014 = 0.984, \tag{17}$$

Thus the basic law of classical probability theory, additivity of probability, and, hence, the formula of total probability, is violated. This violation is a sign that, to describe cell's behaviour, a more complex version of probability theory has to be used. This is the quantum-like probabilistic model corresponding to contextual behaviour. We state again that we are not looking for physical quantum sources of violation of classical probabilistic rules. We couple nonclassical probability with nontrivial contextuality of cell's reactions.

We now can find the coefficient of interference corresponding to the value $\xi = +1$:

$$\lambda_+ = (0.014 - 0.984)/2\sqrt{0.973 \times 0.011} \approx -4.3$$

We see that interference (destructive) is very strong, essentially stronger than typical interference for quantum physical systems. This situation can be described by the hyperbolic probability amplitude:

$$\psi_+ = \sqrt{0.973} - e^{2.138j}\sqrt{0.011} \approx 0.986 - e^{2.138j}0.105,$$

where $\theta_+ = 2.138 = \text{arccosh}\,|\lambda_+| = \text{arccosh}\,4.3$. Then the hyperbolic version of Born's rule, see (13), gives

$$P(\xi = +1|L \cup G) = |\psi_+|^2.$$

We operated with contexts L, the presence of lactose, and G, the presence of glucose, without pointing to concrete levels of concentrations of corresponding molecules. This description is justified by the following remark:

Remark. We recall that lactose induces the enzyme, but without induction certain percentage would be expressed by fluctuation of gene expression. The concentration of glucose is not important, but the following should be taken into account:

If we add 0.2% glucose in the medium, cells can grow to its stationary phase on glucose only and they do not try to utilize lactose. So, if we want to see the enzyme induction during the growth, we have to limit the glucose concentration, mostly usually 0.02%. That amount is insufficient for support of cell growth, then cells try to utilize lactose after consumption of glucose. If we add only 0.02% glucose in the medium without any other carbon (energy) source, then the enzyme level would be similar as in the presence of 0.2% glucose and cells stop growing. If there is any other carbon source than 0.02% glucose, then cells continue to grow and the enzyme level changes depending on the kind of carbon source (for lactose, the level is quite high; for maltose, the level would be low, but significant; for pepton (amino acid mixture), the level would be a little bit more).

4   Decision Making as Decoherence of Quantum-Like Mental State

Dynamical models of decision making are of the main interest for us. We recall that, in a few papers [4], [5], [14], the process of decision making was described by Schrödinger's evolution of the mental state. The latter was assumed to be a pure state (it is represented by a normalized vector of a complex Hilbert space).

In [16] decision making had been represented by more complicated dynamics which describes the evolution of a quantum-like state interacting with an environment. Such dynamics plays an important role in quantum physics. Its fundamental feature is transformation of pure states (described by complex vectors) into mixed states (described by density matrices) – *decoherence.* In our cognitive model memory is an important part of the "mental environment" which induces decoherence of a pure mental state. We have not yet modeled the process of interaction with memory; as often in quantum information theory we represent memory (as well as the external mental environment) as a bath, in our case a "mental bath." In a future paper we plan to model this process in more detail by using the apparatus of quantum Markov chains, cf. [8].

In quantum physics interaction of a quantum system with a bath is described by a quantum version of the master equation. Quantum Markovian dynamics given by the Gorini-Kossakowski-Sudarshan-Lindblad (GKSL) equation, see e.g. [18] for detail, is the most popular approximation of quantum dynamics in the presence of interaction with a bath. We remind shortly the origin of the GKSL-dynamics. The starting point is that the state of a composite system, a quantum system s combined with a bath, is a pure quantum state, complex vector Ψ, which evolution is described by Schrödinger's equation. This is an evolution in a Hilbert space of the huge dimension (since a bath has so many degrees of freedom). The existence of the Schrödinger dynamics in the huge Hilbert space has a merely theoretical value. Observers are interested in the dynamics of the state ϕ_s of the quantum system s. The next fundamental assumption in derivation of the GKSL-equation is the Markovness of the evolution, the absence of long term memory effects. It is assumed that interaction with the bath destroys such effects. Thus, the GKSL-evolution is Markovian evolution. Finally, we point to the condition of the factorizability of the initial state of a composite system (a quantum system coupled with a bath), $\Psi = \phi_s \otimes \phi_{bath}$, where $\otimes$ is the sign of the tensor product. Physically factorization is equivalent to the absence of correlations (at the beginning of evolution; later they are induced by the interaction term of Hamiltonian – the generator of evolution). One of distinguishing features of the evolution under the mentioned assumptions is the existence of one or a few *equilibrium points.* The state of the quantum system s stabilizes to one of such points in the process of evolution; a pure initial state, a complex vector ψ_s, is transformed into a mixed state, a density matrix $\rho_s(t)$. In contrast to the GKSL-evolution, the Schrödinger evolution does not induce stabilization; any solution different from an eigenvector of Hamiltonian will oscillate for ever. Another property of the Schrödinger dynamics is that it always transfers a pure state into a pure state, i.e., a vector into a vector. And we want to obtain mixed states, diagonalized in the basis corresponding to the decision operator. The GKSL-evolution gives such a possibility.

In the process of decision making the brain selects a pure mental state describing possible decisions of the problem under consideration and drives this state. We denote this mental state by ϕ_A. In the process of decision making in games of the Prisoner's Dilemma type (involving two players, Alice and Bob) the state ϕ_A is superposition of possible decisions of Alice in her game with Bob. The state of the mental environment is represented by another complex vector, say ϕ_B. In general, this is a huge mental state representing all superpositions in memory and even permanent supply of superpositions created by the brain through its interaction with the environment. However, if Alice is concentrated on her strategy of gambling with Bob, we can restrict ϕ_B to Alice's mental image of the possible actions of Bob. In reality ϕ_B belongs to complex Hilbert space of a large dimension. Therefore the standard assumption used in the derivation of the GKSL-equation is fulfilled. Nevertheless, we can consider a toy model in which ϕ_B is two dimensional, representing superposition of possible actions of Bob created in Alice's brain. (Thus formally one of the most important assumptions of derivation of the GKSL-equation is not fulfilled. However, more detailed analysis shows that, in fact, in quantum physics the dimension of a bath is not crucial. The crucial property of a bath is that it is very stable to fluctuations in the quantum system s interacting with it. This assumption is fulfilled if Alice's image of possible actions of Bob is sufficiently stable with respect to fluctuations of the state of her possible actions.) The assumption of Markovness of the mental state evolution in decision making is natural. To proceed quickly to a decision, Alice must ignore the history of her reflections on possible actions with respect to Bob.[1] An input from (long-term) memory or mental environment destroys

---

[1] Such reflections are processed in her working memory. So, we discuss Markovness of working memory. Of course, in our model long-term memory is not ignored; it is a part of the mental bath.

(working) memory of her reflections. (Working memory does not preserve a long chain of Alice's reflections.) Finally, we can assume that the initial composite state is factorized, i.e., correlations between Alice's image of Bob and her possible actions are created in the process of decision making. Under these assumptions we can model the process of decision making by using the GKSL-equation.

The mental state representing possible actions of Alice stabilizes to one of equilibrium points of the GKSL-dynamics. (In the mathematical model stabilization is achieved only in the limit $t \to \infty$. However, in reality the brain cannot wait too long. We can assume the presence (in the brain) of a threshold ϵ which is used to terminate the process of stabilization of the mental state to a point of equilibrium.) A model equation considered in this paper has a single equilibrium point. Thus Alice elaborates the unique solution (which depends only on the mental environment, in particular, memory). However, in general the GKSL-equation can have a few different equilibrium points. In such a case depending on the initial state of mind Alice can obtain different solutions of the same problem. Such equations with a richer structure of equilibrium points will be studied in one of coming papers.

Mathematical details of the model of thinking through decoherence can be found in [16].

Acknowledgments. This study was done during visits of I. Basieva and A. Khrennikov to the QBIC (Quantum Bio-Informatics Center, Tokyo University of Science) in 2010 (March, October) and 2011 (March) and visits of M. Asano, M. Ohya, and Y. Tanaka to International Center for Mathematical Modelling in Physics and Cognitive Sciences, Linnaeus University in June 2010 and June 2011.

References

1. Accardi, L., Khrennikov, A., Ohya, M.: The problem of quantum-like representation in economy, cognitive science, and genetics. In.: Quantum Bio-Informatics II: From Quantum Information to Bio-Informatics, pp. 1–8, WSP, Singapore (2008)
2. Khrennikov, A.: Quantum-like Formalism for Cognitive Measurements. Biosystems 70, 211–233 (2003)
3. Khrennikov, A.: On Quantum-like Probabilistic Structure of Mental Information. Open Systems and Information Dynamics 11 (3), 267–275 (2004)
4. Khrennikov, A.: Quantum-like Brain: Interference of Minds. BioSystems 84, 225–241 (2006)
5. Busemeyer, J. B., Wang, Z. and Townsend, J. T.: Quantum Dynamics of Human Decision Making. J. Math. Psychology 50, 220–241 (2006)
6. P. La Mura: Projective Expected Utility. In: Quantum Interaction-2, pp. 87–93, College Publications, London (2008)
7. Franco, R.: The Conjunction Fallacy and Interference Effects. J. Math. Psychol. 53, 415–422 (2009)
8. Accardi, L., Khrennikov, A., Ohya, M.: Quantum Markov Model for Data from. Shafir-Tversky Experiments in Cognitive Psychology. Open Systems and Information Dynamics 16, 371–385 (2009)

9. Khrennikov, A.: Quantum-like Model of Cognitive Decision Making and Information Processing. Biosystems 95, 179–187 (2009)
10. Conte, E., Khrennikov, A., Todarello, O., Federici, A., Zbilut, J. P.: Mental States Follow Quantum Mechanics during Perception and Cognition of Ambiguous Figures. Open Systems and Information Dynamics 16, 1–17 (2009)
11. Khrennikov, A., Haven, E.: Quantum Mechanics and Violations of the Sure-thing Principle: the Use of Probability Interference and other Concepts. J. Math. Psychol. 53, 378–388 (2009)
12. Bruza, P.D., Kitto, K., Nelson, D., McEvoy, C.: Is there something Quantum-like about the Human Mental Lexicon? J. Math. Psychol. 53 362–377 (2009)
13. Lambert- Mogiliansky, A., Zamir, S., and Zwirn, H.: Type Indeterminancy: A model of the KT (Kahneman Tversky) Type Man. J. Math. Psychol. 53 (5) 349–361 (2009)
14. Pothos, E. M., Busemeyer, J. R.: A Quantum Probability Explanation for Violation of Rational Decision Theory. Proc, Royal. Soc. B 276 (1165), 2171–2178 (2009)
15. Khrennikov, A.: Ubiquitous Quantum Structure: from Psychology to Finance. Springer, Heidelberg-Berlin-New York (2010)
16. Asano, M., Khrennikov, A., Ohya, M.: Quantum-Like Model for Decision Making Process in Two Players Game A Non-Kolmogorovian Model. Found. Phys. 41, 538–548 (2010)
17. Inada, T., Kimata, K., Aiba, H.: Mechanism Responsible for Glucose-lactose Diauxie in Escherichia Coli Challenge to the cAMP Model. Genes and Cells 1, 293–301 (1996)
18. Ingarden, R. S., Kossakowski, A., Ohya, M.: Information Dynamics and Open Systems: Classical and Quantum Approach. Kluwer, Dordrecht (1997)

A Quantum-Conceptual Explanation of Violations of Expected Utility in Economics

Diederik Aerts[1], Jan Broekaert[1], Marek Czachor[2], and Bart D'Hooghe[1]

[1] Center Leo Apostel, Brussels Free University
Krijgskundestraat 33, B-1160 Brussels, Belgium
{diraerts,jbroekae,bdhooghe}@vub.ac.be
[2] Katedra Fizyki Teoretycznej i Informatyki Kwantowej,
Politechnika Gdanska, 80-952 Gdansk, Poland
mczachor@pg.gda.pl

Abstract. The expected utility hypothesis is one of the building blocks of classical economic theory and founded on Savage's Sure-Thing Principle. It has been put forward, e.g. by situations such as the Allais and Ellsberg paradoxes, that real-life situations can violate Savage's Sure-Thing Principle and hence also expected utility. We analyze how this violation is connected to the presence of the 'disjunction effect' of decision theory and use our earlier study of this effect in concept theory to put forward an explanation of the violation of Savage's Sure-Thing Principle, namely the presence of 'quantum conceptual thought' next to 'classical logical thought' within a double layer structure of human thought during the decision process. Quantum conceptual thought can be modeled mathematically by the quantum mechanical formalism, which we illustrate by modeling the Hawaii problem situation — a well-known example of the disjunction effect — generated by the entire conceptual landscape surrounding the decision situation.

Keywords: Expected utility, disjunction effect, quantum modeling, quantum conceptual though, ambiguity aversion, concept combinations.

1 Introduction

A basic principle of the von Neumann-Morgenstern theory [24] is Savage's 'Sure-Thing Principle' [22], which is equivalent to the independence axiom of expected utility theory. Over the years, different modified versions of and critiques on von Neuman-Morgenstern's original axiomatization of expected utility have emerged. The Allais paradox [11] and the Ellsberg paradox [16], for example, point to an inconsistency with the predictions of the expected utility hypothesis, indicating a violation of the independence axiom and the Sure-Thing Principle.

In recent works, we have analyzed aspects of human thought [3,7] from the perspective of ongoing investigations on concepts and how they combine, and an approach to use the quantum-mechanical formalism to model such combinations of concepts [6,8,9,18]. In this way, we have shown [3,7] that two superposed layers can be distinguished in human thought: (i) a layer incorporating essentially

D. Song et al. (Eds.): QI 2011, LNCS 7052, pp. 192–198, 2011.

logical thought, (ii) a layer given form under the influence of the surrounding conceptual landscapes each with properties as a whole instead of logically combined sub-concepts. The process in this second layer was labeled 'quantum-conceptual thought' [3,7]. A substantial part of the 'quantum-conceptual thought process' can be modeled by quantum-mechanical probabilistic and mathematical structures. We will look at the violation of the Sure-Thing Principle connected to what psychologists call the disjunction effect [23] and how an explicit quantum-mechanical model for this 'quantum-conceptual thought' can be proposed to describe this type of situation, complementary to approaches presented in the literature [13,14,17,20,21,25].

2   The Sure-Thing Principle and the Disjunction Effect

Savage introduced the Sure-Thing Principle [22] which is equivalent to the independence axiom of expected utility theory: 'independence' meaning 'if subjects are indifferent in their choice between simple lotteries L_1 and L_2, they will also be indifferent in choosing between L_1 mixed with an arbitrary simple lottery L_3 with probability p and L_2 mixed with L_3 with the same probability p.'

The above situation is similar to what in psychology is called the disjunction effect. A well-known example of this disjunction effect is the so-called Hawaii problem [23], which is about the following two situations.

Disjunctive version: Imagine that you have just taken a tough qualifying examination. It is the end of the fall quarter, you feel tired and run-down, and you are *not sure* that you passed the exam. In case you failed you have to take the exam again in a couple of months after the Christmas holidays. You now have an opportunity to buy a very attractive 5-day Christmas vacation package to Hawaii at an exceptionally low price. The special offer expires tomorrow, while the exam grade will not be available until the following day. Would you: x buy the vacation package; y not buy the vacation package; z pay a $5 nonrefundable fee in order to retain the rights to buy the vacation package at the same exceptional price the day after tomorrow after you find out whether or not you passed the exam?

Pass/fail version: [Imagine *...idem...* run-down], and you find out that you *passed* the exam (*failed* the exam. You will have to take it again in a couple of months after the Christmas holidays). [You now *...idem...* after tomorrow].

In the Hawaii problem, more than half of the subjects chose option x (buy the vacation package) if they knew the outcome of the exam (54% in the pass condition and 57% in the fail condition), whereas only 32% did so if they are uncertain about the outcome of the exam.

This Hawaii problem demonstrates clearly a violation of the Sure-Thing Principle triggered by 'uncertainty aversion' (see also[12]). Indeed, subjects prefer option x (to buy the vacation package) when they know that they passed the exam and they also prefer x when they know that they failed the exam, but they refuse x (or prefer z) when they don't know whether they passed or failed the exam.

3   Quantum Modeling of the Hawaii Disjunction Effect

We now put forward an explicit example of a quantum model for the disjunction effect [1,2,3,6,7,8,9], as suggested for the Hawaii problem in [15]. Let us denote by A the conceptual situation in which the subject has *passed* the exam, and by B the *failed* situation. The disjunction of both conceptual situations, i.e. 'A or B', is the conceptual situation where the subject *has passed or failed the exam*.

We represent A by a unit vector $|A\rangle$ and B by a unit vector $|B\rangle$ in a complex Hilbert space $\mathcal{H}$. We take $|A\rangle$ and $|B\rangle$ orthogonal, hence $\langle A|B\rangle = 0$, and describe the disjunction 'A or B' by means of the normalized superposition state $\frac{1}{\sqrt{2}}(|A\rangle + |B\rangle)$. The decision to be made is 'to buy the vacation package' or 'not to buy the vacation package'. This decision is now described by a projection operator M of the Hilbert space $\mathcal{H}$. The probability for an outcome 'yes' (buy the package) in the 'pass' situation (state $|A\rangle$) is 0.54, and let us denote this probability by $\mu(A) = 0.54$. The probability for an outcome 'yes' (buy the package) in the 'fail' situation (state $|B\rangle$) is 0.57, i.e. in our notation $\mu(B) = 0.57$. The probability for an outcome 'yes' (buy the package) in the 'pass or fail' situation (state $\frac{1}{\sqrt{2}}(|A\rangle + |B\rangle)$) is 0.32, i.e. in our notation $\mu(A \text{ or } B) = 0.32$. In accordance with the quantum rules we have

$$\mu(A) = \langle A|M|A\rangle, \ \mu(B) = \langle B|M|B\rangle, \ \mu(A \text{ or } B) = \frac{(\langle A| + \langle B|)}{\sqrt{2}} M \frac{(|A\rangle + |B\rangle)}{\sqrt{2}} \quad (1)$$

Applying the linearity of Hilbert space and taking into account that $\langle B|M|A\rangle^* = \langle A|M|B\rangle$, we have

$$\mu(A \text{ or } B) = \frac{\mu(A) + \mu(B)}{2} + \Re\langle A|M|B\rangle \quad (2)$$

where $\Re\langle A|M|B\rangle$ is the real part of the complex number $\langle A|M|B\rangle$, i.e. the interference term which allows to produce a deviation from the average value.

This 'quantum model based on superposition and interference' can be realized in a three-dimensional complex Hilbert space $\mathbb{C}^3$. (For a more detailed analysis we refer to [1,2,3].) In case (i) $\mu(A) + \mu(B) \leq 1$, we put $a = 1 - \mu(A), b = 1 - \mu(B)$ and $\gamma = \pi$, and in case (ii) $1 < \mu(A) + \mu(B)$, we put $a = \mu(A)$, $b = \mu(B)$ and $\gamma = 0$. We choose

$$|A\rangle = (\sqrt{a}, 0, \sqrt{1-a}) \quad (3)$$

$$|B\rangle = e^{i(\beta+\gamma)}\left(\sqrt{\frac{(1-a)(1-b)}{a}}, \sqrt{\frac{a+b-1}{a}}, -\sqrt{1-b}\right)$$

$$\text{if} \quad a \neq 0; \quad |B\rangle = e^{i\beta}(0,1,0) \quad \text{if} \quad a = 0 \quad (4)$$

$$\beta = \arccos\left(\frac{2\mu(A \text{ or } B) - \mu(A) - \mu(B)}{2\sqrt{(1-a)(1-b)}}\right)$$

$$\text{if} \quad a \neq 1, b \neq 1; \quad \beta \text{ is arbitrary if } a = 1 \text{ or } b = 1 \quad (5)$$

We take $M(\mathbb{C}^3)$ the ray spanned by the vector $(0,0,1)$ in case $\mu(A)+\mu(B) \leq 1$, and we take $M(\mathbb{C}^3)$ the subspace of $\mathbb{C}^3$ spanned by vectors $(1,0,0)$ and $(0,1,0)$ in case $1 < \mu(A) + \mu(B)$. This gives rise to a quantum-mechanical description of the situation with probability weights $\mu(A), \mu(B)$ and $\mu(A \text{ or } B)$. Let us verify this. We have that both vectors $|A\rangle$ and $|B\rangle$ are unit vectors, since $\langle A|A \rangle = a + 1 - a = 1$ and either $\langle B|B \rangle = \frac{(1-a)(1-b)}{a} + \frac{a+b-1}{a} + 1 - b = 1$ in case $a \neq 0$ or $\langle B|B \rangle = 1$ trivially in case $a = 0$. For both cases of a, one can easily check that $\langle A|B \rangle = 0$, e.g. $\langle A|B \rangle = \sqrt{(1-a)(1-b)}e^{i(\beta+\gamma)} - \sqrt{(1-a)(1-b)}e^{i(\beta+\gamma)} = 0$ for $a \neq 0$, which shows that $|A\rangle$ and $|B\rangle$ are orthogonal. Now we only need to check whether this model yields the correct probabilities in the expressions (1).

First, let us consider $a \neq 0, a \neq 1, b \neq 1$. In case that $\mu(A) + \mu(B) \leq 1$, we have $\langle A|M|A \rangle = 1 - a = \mu(A)$, $\langle B|M|B \rangle = 1 - b = \mu(B)$, and $\langle A|M|B \rangle = -\sqrt{(1-a)(1-b)}e^{i(\beta+\gamma)} = \sqrt{(1-a)(1-b)}e^{i\beta}$. In case $1 < \mu(A) + \mu(B)$, we have $\langle A|M|A \rangle = a = \mu(A)$, $\langle B|M|B \rangle = \frac{(1-a)(1-b)}{a} + \frac{a+b-1}{a} = \frac{ab}{a} = b = \mu(B)$, and $\langle A|M|B \rangle = \sqrt{a}\sqrt{\frac{(1-a)(1-b)}{a}}e^{i\beta} = \sqrt{(1-a)(1-b)}e^{i\beta}$. Hence in both cases we have $\Re\langle A|M|B \rangle = \sqrt{(1-a)(1-b)}\cos\beta$, so that $\Re\langle A|M|B \rangle = \frac{1}{2}(2\mu(A \text{ or } B) - \mu(A) - \mu(B))$. Applying (5) this gives $\mu(A \text{ or } B) = \frac{1}{2}(\mu(A) + \mu(B)) + \Re\langle A|M|B \rangle$, which corresponds to (2). This shows that, given the values of $\mu(A)$ and $\mu(B)$, the correct value for $\mu(A \text{ or } B)$ is obtained in this quantum-model representation.

In the present Hawaii problem we have $\mu(A) = 0.54$, $\mu(B) = 0.57$ and $\mu(A \text{ or } B) = 0.32$. First, let us note that this means a classical model is not allowed, since $\mu(A \text{ or } B) < \mu(B)$. Since we have $1 < \mu(A)+\mu(B) = 1.11$, we put $a = 0.54$, $b = 0.57$ and $\gamma = 0$, and we take $M(\mathbb{C}^3)$ the subspace of $\mathbb{C}^3$ spanned by vectors $(1,0,0)$ and $(0,1,0)$. Finally, according to equations (3), (4) and (5), we obtain $|A\rangle = (0.7348, 0, 0.6782)$, $|B\rangle = e^{i121.8967°}(0.6052, 0.4513, -0.6557)$. In [3] similar vectors and angles for a number of experimental data in concept theory have been calculated. For some of these items $\mathbb{C}^3$ models do not exist, requiring to extend the modeling to Fock space [2].

4   Concept Combinations, The Disjunction Effect and Conceptual Landscapes

The disjunction effect, apparent in 'decision theory', was modelled using quantum game theory [13,21], and quantum theoretical models [14,20,25] along similar lines as our previous model [1,2,3,4]. In different terms the effect was studied experimentally in problems occurring with the combination of concepts [19]. In this section we give an example of concept disjunction which reveils 'overextension' in contrast to 'underextension' as in the Hawaii problem [19], and explain how the 'inverse disjunction effect' for concept combinations can be understood as being due to the presence of what we have called quantum conceptual thought. The pair of concepts *Fruits* and *Vegetables* and their disjunction *Fruits or Vegetables*, gives for the item *Olive* the membership weights 0.5, 0.1 and 0.8 related to *Fruits*, *Vegetables* and *Fruits or Vegetables* respectively. We can prove that for these weights it is not possible to find a Kolmogorovian representation [3]. This means that these weights cannot be obtained by supposing that subjects reasoned following classical logic and that the weights are the result of a lack of knowledge about the exact outcomes given by each of the individual subjects.

Indeed, if 50% of the subjects have classified the item *Olive* as belonging to Fruits, and 10% have classified it as belonging to *Vegetables*, then following classical reasoning at most 60% of the subjects can classify it as belonging to '*Fruits or Vegetables*', while the experiment shows that 80% did so. This means that these weights arise in a distinct way. Some individual subjects must necessarily have chosen *Olive* as a member of '*Fruits or Vegetables* and 'not as a member' of *Fruits* and also 'not as a member' of *Vegetables*, otherwise the weights 0.5, 0.1 and 0.8 would be impossible results. Concretely, this means that for the item *Olive*, the subject considers '*Fruits or Vegetables*' as a newly emerging concept and not as a classical logical disjunction of the two concepts *Fruits* and *Vegetables* apart. In this 'quantum-conceptual' thought process the emergence of a new concept '*Fruits or Vegetables*', within the landscape of existing concepts, i.e. *Fruits*, *Vegetables* and *Olive*, gives rise to the deviation from the membership weight expected from classical logic (0.8 is strictly bigger than 0.5 + 0.1).

Is it possible to apply quantum-conceptual thought in the disjunction of concepts to explain the traditional disjunction effect in the Hawaii problem? There is a set of experiments [12], although performed with a different goal, which confirm that our explanation for concepts and their disjunction is also valid for the traditional disjunction effect. These experiments reconsider the Hawaii problem to show that the disjunction effect does not depend on the presence of uncertainty (pass/fail) but on the introduction into the text-problem of a non-relevant goal [12]. This indicates in a very explicit way that it is the overall conceptual landscape that gives form to the disjunction effect. More specifically, the authors point out that option z contains an unnecessary goal, i.e. that one needs to 'pay to know', which is independent of the uncertainty condition. In this sense, their hypothesis is that the choice of option z occurs as a consequence of the construction of the discourse problem itself [12]. Option z is not a real alternative to x and y, but becomes an additional premise that conveys information which changes the decisional conceptual landscape. These results support the view that the disjunction effect appears when a suitable decisional conceptual landscape is present rather than mainly depending on the presence of uncertainty.

5   Conclusion

In our earlier work [3,7] we introduced the notion of 'conceptual landscape' as a natural extension of our approach in the modeling of concepts and their combinations [6,8,9,18]. We demonstrated that in decisions the 'entire' conceptual landscape should be taken into account and modeled within our quantum modeling scheme, implicitly inducing the subject's notion of 'worldview' [5,10]. All elements of a subject's worldview surrounding a given situation which can possibly influence a human decision should be taken into account, i.e. if these can be expressed conceptually. These elements can then be taken into account by means of the quantum modeling scheme we have developed in earlier work for concepts and their combinations [1,2,3,7,8,9,18]. This being the case, we are already able to grasp a very important aspect and also fundamental part of the dynamics generated by the totality of the worldview influence.

References

1. Aerts, D.: Quantum interference and superposition in cognition: Development of a theory for the disjunction of concepts (2007a), Archive Reference and Link: http://arxiv.org/abs/0705.0975
2. Aerts, D.: General quantum modeling of combining concepts: A quantum field model in Fock space (2007b), Archive reference and link: http://arxiv.org/abs/0705.1740
3. Aerts, D.: Quantum structure in cognition. J. Math. Psy. 53, 314–348 (2009)
4. Aerts, D., Aerts, S.: Applications of quantum statistics in psychological studies of decision processes. Foundations of Science 1, 85–97 (1994)
5. Aerts, D., Apostel, L., De Moor, B., Hellemans, S., Maex, E., Van Belle, H., Van der Veken, J.: Worldviews, from Fragmentation towards Integration. VUBPress (1994)
6. Aerts, D., Broekaert, J., Gabora, L.: A case for applying an abstracted quantum formalism to cognition. New Ideas in Psychology 29, 136–146 (2010)
7. Aerts, D., D'Hooghe, B.: Classical logical versus quantum conceptual thought: Examples in economics, decision theory and concept theory. In: Bruza, P., Sofge, D., Lawless, W., van Rijsbergen, K., Klusch, M. (eds.) QI 2009. LNCS, vol. 5494, pp. 128–142. Springer, Heidelberg (2009)
8. Aerts, D., Gabora, L.: A theory of concepts and their combinations I: The structure of the sets of contexts and properties. Kybernetes 34, 167–191 (2005a)
9. Aerts, D., Gabora, L.: A theory of concepts and their combinations II: A Hilbert space representation. Kybernetes 34, 192–221 (2005b)
10. Aerts, D., Van Belle, H., Van der Veken, J. (eds.): Worldviews and the Problem of Synthesis. Springer, Dordrecht (1999)
11. Allais, M.: Le comportement de l'homme rationnel devant le risque: critique des postulats et axiomes de l'école Américaine. Econometrica 21, 503–546 (1953)
12. Bagassi, M., Macchi, L.: The 'vanishing' of the disjunction effect by sensible procrastination. Mind & Society 6, 41–52 (2007)
13. Busemeyer, J.R., Wang, Z., Townsend, J.T.: Quantum dynamics of human decision-making. Journal of Mathematical Psychology 50, 220–241 (2006)
14. Busemeyer, J.R., Pothos, E., Franco, R., Trueblood, J.: A quantum theoretical explanation for probability judgment 'errors'. Psychological Review 118(2), 193–218 (2011)
15. Danilov, V.I., Lambert-Mogiliansky, A.: Measurable systems and behavioral sciences. Mathematical Social Sciences 55(3), 315–340 (2008)
16. Ellsberg, D.: Risk, ambiguity, and the Savage axioms. Quarterly Journal of Economics 75(4), 643–669 (1961)
17. Franco, R.: Risk, Ambiguity and Quantum Decision Theory (2007), Archive reference and link: http://arxiv.org/abs/0711.0886
18. Gabora, L., Aerts, D.: Contextualizing concepts using a mathematical generalization of the quantum formalism. Journal of Experimental and Theoretical Artificial Intelligence 14, 327–358 (2002)

19. Hampton, J.A.: Disjunction of natural concepts. Memory & Cognition 16, 579–591 (1988)
20. Khrennikov, A.: Quantum-like model of cognitive decision making and information processing. Biosystems 95, 179–187 (2008)
21. Pothos, E.M., Busemeyer, J.R.: A quantum probability explanation for violations of 'rational' decision theory. Proceedings of the Royal Society B (2009)
22. Savage, L.J.: The Foundations of Statistics. Wiley, New-York (1954)
23. Tversky, A., Shafir, E.: The disjunction effect in choice under uncertainty. Psychological Science 3, 305–309 (1992)
24. von Neumann, J., Morgenstern, O.: Theory of Games and Economic Behavior. Princeton University Press, Princeton (1944)
25. Yukalov, V.I., Sornette, D.: Decision theory with prospect interference and entanglement. Theory and Decision 70, 283–328 (2010)

On the Nature of the Human Mind: The Cognit Space Theory

George Economides

St Peter's College, University of Oxford, New Inn Hall Street, OX1 2DL, UK
george.economides@spc.ox.ac.uk

Abstract. The Cognit Space theory of how the human mind works is presented. A new version of the "Two Minds" hypothesis is introduced, separating the Human Evolutionary Adapted Mind (HEAM) from the Tabula Rasa Mind (TRM). Consciousness is suggested to be the real time optimisation of a mental state wavepacket with respect to a person's value system.

Keywords: quantum cognition, two mind, mental states, consciousness, split-brain, sleep, cognit, concepts, dual processing.

1   Introduction

This paper combines evolutionary psychology, cognitive neuroscience and the psychological hypothesis of "Two Minds" [4][6] with concepts of quantum dynamics to formulate a falsifiable theory of how the human mind works. In this work the mind is examined as an emergent property of the higher functions of the brain[2].

2   The Two Mind Distinction

The idea of "Two Minds" has long been supported by a community of psychologists[4], yet there is no universal agreement on how these are related, distinguished or interacting[6]. In the Cognit Space theory, the mind is separated into two parts, by arguing that if the rate of change of one part is comparable to the course of evolution, then this can be distinguished from another part for which the rate of change is comparable to the lifetime of a person. The former part is named the *Human Evolutionarily Adapted Mind* (HEAM) while the latter the *Tabula Rasa Mind* (TRM).

The TRM is a tool for non-random adaptation to the present and expected environments. This dedicated part of the mind is future-oriented, aims to understand reality and predict the selective advantages[1]. Coupled to this idea is an argument used by evolutionary psychologists: certain pleasurable[2] experiences

---

[1] Of course, these "adaptations" do not alter the genetic code.

[2] Here defined as a continuous pleasure-pain axis: more pleasure is synonymous with less pain. Also, what is more pleasurable is more desirable.

D. Song et al. (Eds.): QI 2011, LNCS 7052, pp. 199–204, 2011.

to a person are so for an evolutionary reason[1]. Thus, when in the evolutionary perspective (an *allocentric* map) the person moves towards what is evolutionarily advantageous, in a personal perspective (an *egocentric* map), he moves towards what is pleasurable. Progressing that argument, the TRM generates a force for the individual to adapt as to maximise present and future instances of pleasure. Since the adaptation is to an unknown environment, this part of the mind would also need to establish what is "beneficial" or "pleasurable".

The HEAM is expected to be evolutionarily older which should be reflected in the brain part that hosts it, here suggested to be sub-cortical structures. The TRM is evolutionarily more recent, which is reflected in the brain region hosting it, the cortex[8].

Unlike previous theories[4] of two minds, here the *"new"* mind is not characterised slow or sequential and although its response might be more complex and thus sometimes slower, in total it is proposed that it has greater processing power[6]. Furthermore, its architecture is considered to be object-oriented. The limits of the abilities of the TRM are part of the HEAM.

3   The Human Evolutionarily Adapted Mind (HEAM)

HEAM is the collection of genetically-deterministic instinctual behaviours that have been shaped by the course of evolution and are present in modern humans, although these might not be activated or used. As evolutionary psychologists point out[1], the time elapsed since the dawn of civilisation is minute on the scale of evolution, and therefore the effects of modern society on HEAM are expected to be equally small.

The behavioural responses of HEAM to the environment are expressed in direct bodily expressions such as emotional and hormonal state alterations. HEAM is an emotional/instinctual mind, and although people may become conscious of it, it does not contain consciousness itself. The stronger the emotional significance of a memory, the more intensely it is recorded in the amygdala[8][5], which is a demonstration that the value system of HEAM is predefined. Combining the emotional response with the episodic memory, the HEAM expresses instinctual behaviour in habitual or procedural ways, which are stored and recalled from sub-cortical structures. This is consistent with the view that the basal ganglia are important in reward-based and conditioned learning and linking actions[5][8][9][3]. The HEAM is always in first person perspective and not sophisticated enough to construct allocentric maps of reality. Moreover, it is present-oriented; concerned about homeostasis.

4   The Tabula Rasa Mind (TRM)

The purpose of the TRM is to model reality and use that model to predict what the best strategies for competitive success are. This is referred to as solving the

---

[3] The TRM only models HEAM as far as it is necessary; otherwise it is content to receive straight output.

Problem of Reality (PoR). A model of reality[4] is a way of functionally representing reality, and is a specific way of connecting reality-elements, or reality-representations. Since modelling the world is part of the nature of the TRM, it is instinctual for humans. In addition, as all elements of reality are defined by their inter-connections, no information is held *in vacuo* but it is linked both conceptually and episodically. Furthermore, this model includes the limitations (or *boundary conditions*) of reality: what is possible and what are distinct concepts. Such a model does not need to be separately stored, if it is included in the properties of the reality elements.

Each reality-element (called *cognit*, D [2] [19])[5] is composed of related projections, p_i, where each projection is an "i-th" value in an axis of quantifiable change ($D = \cup_{i,j} p_{i,j}$). Different projections, p_i, can be thought of as different ways that sensory stimuli are recorded, e.g. different senses, therefore appearing at the corresponding part of the cortex[6]. Shared projections link different cognits. Therefore, a cognit of X is the space of all possible "X", and itself is amodal[7]. The cognit of a person for himself is called the *Ego-Cognit* (EC).

It is accepted that conceptual categorization occurs for auditory and visual stimuli[5][8], it is here argued that cognits themselves are discrete and thus create a quantised space which hosts the model of reality. Since the solutions of the PoR occur in a quantised confined space, they themselves are quantised, and they can be expressed in terms of cognit connections. These solutions are named the stationary mental states of the individual (ω).

Mathematically, each cognit may be treated as a tensor, whose rank would depend on the number of projections it has. Each projection itself is a tensor, which in the case of language, it is in agreement with modern computational semantics. The similarity between different cognits can be quantified in terms of their mathematical similarity, as $\sqrt{\frac{(D_1 \cdot D_2)^2}{(D_1 \cdot D_1) \cdot (D_2 \cdot D_2)}}$. Cognit similarity may be thought of in terms of their meaning overlap (e.g. hand and leg) and granularity.

---

[4] Early in the life of a person, a universal model might not be possible so one makes local (with respect to environmental variables) models, but there is a natural tendency to link them.

[5] In contrast to Schnelle's[2] and Fuster's[19] theories, here cognits are thought of as only localised in the cortex and being able to form linguistic projections ("LF-Cognits"), "M-Cognits" are not discreet and would not give rise to quantised space.

[6] For example, seeing a table creates an instance (ϵ) of "table" in the visual projection space (an instance in total is $\epsilon_j = \cup_i p_{i,j}$). A projection may be subdivided to different axes, each corresponding to a brain area that distinguishes characteristics that make this projection unique[5][8][2]. Alternatively, simultaneously mapping a visual stimulus on the different axes that the brain uses (in units of neuronal excitation) would give one instance of visual projection of the cognit "table", $p_{Visual,1}^{Table}$. Similarly, every image of a table gives a different instance: $p_{Visual,2}^{Table}, p_{Visual,3}^{Table}$, etc. Projections themselves are neuronal assemblies, and they might overlap [19]. In [10], projections are "*assembly connection*", while cognits give rise to "*convergence connection*".

[7] Moreover, the *core* of the cognit can be defined, D^c which is the characteristics of the cognit that are common to all instances ($D^c = \cap_j \epsilon_j$).

The meaning of a cognit is altered every time is it re-defined, e.g. in order to link it to more projections[8].

The answer to possible cognit overpopulation, is based on the phenomenon of plasticity, and it is named the *Economy principle*: *cognits and projections become or remain distinct if that distinction is practically useful to the individual*. If the Economy principle is not satisfied, similar cognits or projections are merged[9]. The flexibility of a cognit to form new connections (interpretations) can be seen as a relative uncertainty. Old neuronal connections (from cognits to projections and to each other) become stronger with use[14] and are less likely to be modified in future models, which contributes to continuity between successive models of reality.

The cognits form one kind of memory, which is linked with HEAM's memory. The model of reality is a best-fit of cognits to empirical observations, and the weight of each observation signifies its emotional importance.

Another aspect of PoR regards what is, and what should be pleasurable. To model that, the TRM forms a hierarchical *Value System* (VS). The initial VS of what is pleasurable arises from the HEAM[10], but once language is acquired, more complex definitions of pleasure and virtue result through social interactions. The pleasure-pain is one more axis in the sense that pain is one more of the senses.

Apart from cognits, the second kind of neuronal connection is called *connectors*. Connectors are input and output pathways with the rest of the body that are formed and refined during the lifetime of a person, on the basis of initial soft-wiring. In the case of output, connectors operate in (sometimes parallel) hierarchical layers that [13] transform what is meant, to something that can be said or done, to muscle movement. Connectors for similar pathways that have a common origin may have shared levels as they are gradually distinguished according to need. The cortical position of connectors affects the hemispheric laterization of cognits and projections. The third kind of neuronal connectivity is called *operators*, which correspond to interactions.

For a given situation that is described by a person's model of reality, the sensory input is transformed via a series of connectors to projection instances, activating specific cognits and the combination of cognits is associated with and thus elicits (in a simplified case a single) mental state. If this is not the optimum in the pleasure/pain axis in the current state, then there is a tendency towards a situation that will be the optimum, or to change to a more pleasurable state. Either of these occurs via the application of an operator. The application of an operator produces output, which will then be transformed by connectors.

---

[8] The cognit space is personal and formed by each individual throughout his life, thus forming a kind of idiolect (an *I-language*) in his attempt to model the external world (an *E-language*) [7].

[9] The application of the Economy principle is modified by the brain's ability to exhibit plasticity.

[10] The TRM has shaped responses that anticipate and modify HEAM reactions, and when these are not given malfunctions may occur[18].

As it has been reported, excited neurons also synchronise their firing rate, and that synchronisation elicits a long-range cognit connection at a specific firing frequency[11][5][10]. As environment stimuli reach the thalamus[12], this gives the rhythm that elicits mental states automatically[8].

5   Consciousness, Sleep and the Silent Man

In reality, a person will not form a reality model precise enough as to have a specific mental state excited for a set of environmental stimuli. Instead, a plethora of mental states will be excited, and therefore, the total state of the individual, $|c\rangle$, may be defined as a wavepacket ($|c\rangle = \sum_i \alpha_i |\omega_i\rangle$). Each mental state would lead to an alternative behaviour and the corresponding probability of that behaviour is proportional to the modulus of the coefficient α_i.

Consciousness is here regarded as the real time optimisation of this wavepacket with respect to the VS. Consciousness thus would be focused to novel characteristics of the environment and with characteristics that may influence significantly the possibility of the individual to succeed in his VS[13]. As consciousness is part of TRM, its computational power is respectively a subset of the computational power of the TRM. The individuality of each person is contained in the cognits, VS and his model of reality.

Consciousness is in close co-operation with the HEAM, which contributes the emotional significance of the events as they happen via a direct route[14]. The co-operation of the TRM with the HEAM may only occur from within the same frame of reference and as HEAM operates from an egocentric perspective, while TRM from an amodal space, one needs to be transformed. Only the stationary states are known to TRM, so the only possible systematic transformation means that consciousness operates from a first person perspective. That also means that consciousness operates at the junction of HEAM and TRM[18], which gives rise to experiencing stimuli in the manner responded to by both minds, the qualia.

During sleep there is the opportunity to modify the model of reality and to experiment with different interactions, without significant risk[15]. Refitting of the model during sleep[16] agrees with the observation that memories are solidified then[8], but also that learning is impaired by sleep deprivation only if the task concerns a new behavioural strategy[12]. It also follows that habitual learning does not depend on sleep[11][15].

---

[11] It is also possible to have mental states that are not environmentally activated, but to result from previous mental states.

[12] Mental states arise from the third kind of connection of [10]: by "*synchrony*".

[13] If there are no environmental stimuli significant enough, then consciousness may experiment with more distant future outcomes.

[14] Either mind might generate unconscious output, so heuristics might arise from either HEAM or TRM.

[15] The cognit space is also modified while awake due to plasticity, but to a lesser degree.

[16] If an event is not fitted in the model of reality but is of great emotional significance, it would be dominant in dreams, itself or by associations.

During sleep, the stimuli are re-introduced in the cortex, but this time, instead of consciousness damping some of the excited mental states, the plasticity increases [8][17]: the cognit space changes as to incorporate the sensory stimuli. By changing the cognit space, the mental states that form standing waves in it change, so new mental states are shaped.

So far, no distinction has been made with respect to the two hemispheres, since regarding them as separate would fragment the cognit space. Yet, due to their high symmetry and the relatively small area of the corpus callosum, the wavepacket of the current state may be divided into two commuting but not identical parts, which would be useful in cases of multitasking[16], although the overall state would remain a single one. Moreover, completely separating the two hemispheres, would result in two wavepackets each confined into a different hemisphere. This is argued to be the case for commissurotomy patients [8][5][3].

6   Conclusions

The outline of the Cognit Space theory has been given. This is the first theory that bridges evolutionary psychology, neuroscience, the psychological hypothesis of Two-Minds and concepts of quantum mechanics and offers an explanation to the phenomena of sleep, consciousness and commissurotomy patients with consistent reasoning. The author thanks the Alexander Onassis Public Benefit Foundation of Greece for funding.

References

1. Dubar, R., Barrett, L., Lycett, J.: Evolutionary Psychology. Oneworld Publications, Oxford (2005)
2. Schnelle, H.: Language in the Brain. CUP, NY (2010)
3. Ramachandran, V.S.: The Emerging Mind. Profile Books Ltd., London (2010)
4. Evans, St.B.T.: Thinking Twice. OUP, NY (2010)
5. Ward, J.: The Student's Guide to Cognitive Neuroscience. Psychology Press, Hamshire (2008)
6. Evans, J.St.B.T., Frankish, K. (eds.): In Two Minds. OUP, NY (2009)
7. Chomsky, N.: On Language. New Press, NY (2007)
8. Bear, M.F., Connors, B.W., Paradiso, M.A.: Neuroscience. Lippincott Williams & Wilkins, USA (2001)
9. Dalgleish, T.: Nature 5, 582 (2004)
10. Mashour, G.A.: NeuroQuantology (I), 29 (2004)
11. Fenn, K., Nusmaum, C.H., Margoliash, D.: Nature 425, 614
12. Maquet, P.: Science 294, 1048 (2001)
13. Scott, S.K., Johnsrude, I.S.: Trends in Neurosciences 26(2), 100 (2003)
14. Edelman, G.: The Remembered Present. Basic Books, N.Y (1989)
15. Wamsley, E.J., Perry, K., Djonlagic, I., Babkes Reaven, L., Stickgold, R.: Sleep 33(1), 59 (2010)
16. Charron, S., Koechlin, E.: Science, 360 (April 16, 2010)
17. Hobson, A., Pace-Schott, E.: Nature Reviews 3, 679 (2002)
18. Damasio, A.: Descarte's Error. Vintage Publications, London (2006)
19. Fuster, J.M.: International Journal of Psychophysiology 60, 125–132 (2006)

Quantum Phenomenology and Dynamic Co-emergence

Christian Flender

University of Freiburg, Germany
flender@iig.uni-freiburg.de

Abstract. Conceptual similarities between phenomenological descriptions of conscious experience and non-local effects as found in quantum mechanics are difficult to dismiss. Our engaged being-in-the-world, for instance when being immersed in reading, writing, or speaking, lacks a clear self-other distinction and mind-body separation as much as combined quantum systems lack separability of entangled states. Our sense of affordances or possibilities, for instance when choosing among several opportunities for action, is strongly reminiscent of superpositions or potentiality states as opposed to the factual reality of eigenstates. Since we can hardly give causally necessary and sufficient conditions for our actions causality breaks as much for intentional action as for state reduction, or wave function collapse, in the quantum world. Intentional action is always already entangled and therefore emerges from embodied and embedded comportment as much as intentionality modulates or submerges our involvement in the world. It is argued that understanding skilful coping as a mode of being-in-the-world is best conceptualized as a dynamically co-emerging whole prior to any mind-body and self-other distinction. Some elements of work practices in air traffic control are discussed as an illustrative application.

1 Introduction

A phenomenology of everyday experience like writing reports, listening to speech, or giving talks, is often taken for granted or artificially disturbed when examined analytically. As many phenomenologists have argued, a Newtonian-Cartesian stance on reality often distorts the actual phenomenon at hand [1]. For instance, the way I ride a bike doesn't require me to represent conditions of satisfaction in order to evaluate my activity. My experience of bike-riding is a unified, engaged and holistic exercise, a mode of being that Heidegger denotes as readiness-to-hand (Zuhandenheit) [2]. This attitude of involvement appears far from being analytically comprehendible. Adopting a detached observer's point of view, that is becoming someone who is merely looking at bike-riding from a theoretical perspective, contrasts with ones immersed activity and skilful coping and makes one enter the realm of analytical thinking. However, unless a breakdown occurs, e.g., I might encounter a punctured tire, there is no reason to confuse our engaged dealings with equipment, e.g., bikes, and the disinterested and wondering mode called presence-at-hand (Vorhandenheit) [2].

D. Song et al. (Eds.): QI 2011, LNCS 7052, pp. 205–210, 2011.

At first sight it turns out to be challenging to subject skilful coping to causal explanation. Accordingly, phenomenology is often considered as purely descriptive and so its value for explanatory purposes is limited. However, first-person experience is a condition for the possibility of any kind of explanation and therefore the modes, or ways, things are given in experience need to be understood before objectification.

Aims of this essay are twofold. First, drawing from recent discussions in the philosophy of mind, it aims to introduce a phenomenological way for looking at work practices. Some important elements of social dynamics get lost by adopting an unexamined reductive or dualist stance on work routines. Such elements, however, are crucial for understanding concerns of certain stakeholders literally involved in work practices. By means of an example in air traffic control phenomenology intimates perspectives of controllers being concerned about changing their work practices, which don't come to the fore in classical viewpoints. To make controllers perspectives more intelligible, that is the second aim of this essay, I argue that quantum theory, in particular the notion of dynamic co-emergence derived from quantum concepts, and phenomenology are mutually enlightening. Not unlike Heidegger, who introduced neologisms, that is, he invented words which will in virtue of their originality be free of any philosophical baggage, I borrow notions from quantum theory which bear the potential to improve intelligibility and so provoke readers to thoughtfulness rather than provide them with simple answers to well-defined problems.

After a short discussion of monism and dualism in the next section, in Section 3, the concept of dynamic co-emergence puts forward one way of looking at the relation between subjective experience and objective features. The quantum effects used to introduce dynamic co-emergence will then be applied for describing viewpoints of work practices in air traffic control.

2   Monism and Dualism

As theorists we can hardly deny philosophical assumptions guiding our endeavours. Some authors have argued that there is no philosophy-free science, there is only science whose philosophical baggage is taken on board without examination [3]. This section briely examines assumptions guiding research to understand actions of stakeholders involved in work practices.

Dualism denotes the view that there are two fundamental realms of being. Substance dualism acknowledges two metaphysical essences, the physical substance as posited by scientific theories and a mental realm of psychological phenomena [4]. In contrast, property dualism commits to one essence, material, ideal or neutral; however, it acknowledges mental and physical properties as separate features. Generally, dualism explicitly separates subjective phenomena (mind) from matter and energy (body). However, it leaves open how each realm causes the other. For instance, if someone stands on my toe this causes me to feel pain. Vice versa, if I consciously choose to raise my arm this causes my body to move. Accordingly, the fundamental problem of dualism is that at present there is no explanation of how mind and matter cause each other. There are many functional explanations of brain processes correlating with mental phenomena. For instance, neuronal activity in the visual cortex correlates with visual experience. However, there are no necessary and sufficient conditions for brain processes to cause conscious experience as much as there are no necessary and sufficient conditions for mental effort to cause physical change. Therefore, dualism is left with what came to be known as the explanatory gap or the hard problem of consciousness [5].

Monism presupposes only one metaphysical realm or substance. For instance, scientific realism and metaphysical idealism are two widely adopted forms of monism. For idealists the whole universe is psychological. According to idealism, what we think of as physical is just one of the forms that the underlying mental reality takes. I do not consider metaphysical idealism here since I believe that the denial of an objective material substrate doesn't contribute to a (dis-)solution of the mind-body problem. Instead, I focus on two other forms of monism, namely behaviourism and functionalism.

Behaviourism is the paradigm for studying causal laws governing the human mind in terms of input-output systems [6]. According to behaviourism, neither physical nor mental processes explain the mind. Rather the causal laws of human behaviour are claimed to be externally observable patterns of stimulus and response. However, behaviourism denies internal states like feelings, hopes, desires, and fears. The need to take internal states into account led to several versions of functionalism. Functionalism is another form of monism underlying attempts to understand the human mind.

Functional models of the mind draw from external stimuli and internal states and so they overcome behaviourism [7]. Moreover, functionalism distinguishes type-identity and token-identity. The former claims that types of mental processes are identical with types of physical processes. The latter holds the view that instances of mental phenomena are identical with instances of physical phenomena. Token-identity is usually associated with supervenience or multiple realizability [8]. For instance, pain can be realized by multiple brain processes though it is identical with patterns of neuronal firings. However, functionalism raises the problem of mental causation [9]. Firstly, minds can cause physical things. I can raise my arm and it moves. Secondly, under the assumption that only physical things can cause physical things, i.e., the physical world is causally closed, minds must be physical. But minds cannot be reduced to the physical due to their undeniable qualitative and subjective feel. Therefore, minds cannot cause physical things. The next section attempts to dissolve this contradiction by introducing quantum phenomenology and the concept of dynamic co-emergence.

3   Quantum Phenomenology: Being-in-the-World

For the purpose of describing actions in phenomenological terms reducing concepts to a material or immaterial substrate is as much problematic as accepting dualism. However, there are several alternatives for defining relations between the mental and the physical without falling into the trap of reductionism and the explanatory gap. The notion of emergence is at the centre of such attempts. In order to distinguish relations between different levels of descriptions, assumptions about necessary and sufficient conditions provide a useful classifcation scheme [10]. Given two levels of description, let's say mind and body, the material level (A) can be related to the psychological description (B) in the following ways.

1. A provides necessary and sufficient conditions to derive B.
This view is called reductionism and implies a form of monism, e.g., scientific realism or metaphysical idealism. For instance, materialists could claim that colour experience in B is exhaustively determined by neurobiological processes in A. Vice versa, idealists might argue that our objectification of neurobiological processes in A is totally determined by our mental ability to reason in B.

2. A provides neither necessary nor sufficient conditions to derive B.
This view is called radical or ontological emergence. There are no determinative relationships between A and B. Furthermore, it implies dualism, i.e., there are two separated levels of being A and B. However, ontological emergence of inherent high-level properties with causal powers is witnessed nowhere [11]

3. A provides necessary but not sufficient conditions to derive B.
This view is called contextual emergence [10]. In order for A to be sufficient contingent conditions have to be introduced in B. Such conditions are contextual or situated. Contextuality requires a distinction between actuality and potentiality. In quantum theory potentiality is modelled as an implicit but not totally realized set of choices. Contingent factors modelled as superposition states are understood as affordances, i.e. affordances to measure or act.

4. A provides sufficient but not necessary conditions to derive B.
This view is called supervenience or multiple realizability [8]. For instance, token identity might assume that instances of mental phenomena are identical with instances of physical phenomena. Therefore, mental states can be multiply instantiated. A headache is a pain which may be realized by multiple brain processes though it is identical with patterns of neuronal firings.

Reductionism (1) and ontological emergence (2) are not viable options for conceptualizing a phenomenology of being-in-the-world. As discussed in the previous section, the former paired with materialism maintains the problem of mental causation; the latter is typical of dualism and therefore it accepts the explanatory gap. However, contextual emergence (3) and supervenience (4) can transcend both extremes and therefore avoid postulating the causal inefficiacy of the mental and the explanatory gap. Paired with a topological constraint where higher level properties in B or global patterns of behaviour enslave or constrain material components in A towards a direction, (3) and (4) make up the concept of dynamic co-emergence (see [12] for a different introduction).

In summary, embodied in the concept of dynamic co-emergence there are four conceptual characteristics shared by quantum theory and phenomenology (see also [13] for a similar approach). (A) A priori, there is no necessary and sufficient condition for something to happen, neither physical nor mental. A being-in-the-world comes into being out of itself. Neither do mental phenomena fully determine and thus cause physical phenomena nor vice versa. A being dynamically co-emerges and so moves as a unified and continuous whole (Indeterminism). (B) For sufficiency, contingent conditions always already extend into the world. In quantum terms when actualizing possibilities time and space are extended. There is neither a sequential ordering of events nor is there locality (Extension). (C) Contextual conditions stand side-by-side and thus they constitute a space of possibilities. According to the superposition principle potentiality states afford to become actualized through action and decision making (Potentiality). (D) In skilful coping subject and object are entangled. The state of an object system prior to measurement is entangled with its environment including observer. After observation the object system is separated from the observer and a superposition state is transformed into a classical state (Non-separability).

In the following I adopt these concepts for examining work practices in air traffic control.

4   A Pertinent Example: Being-in-the-Air

All around the world air traffic centres manage aircrafts, their coming and going, ascending and descending. A recent publication presents the results of a comparative study of eight different control centres in France and the Netherlands [14]. A main inspiration for a careful ethnographic study was the fact that many attempts to automate air traffic control have failed. This was mainly due to controllers who remained attached to a paper strip.

Paper strips are shared among controllers and annotated for several purposes. Generally, they symbolize aircrafts passing through sectors into which air space is divided. Several minutes before an aircraft enters a sector, a strip is printed and collected by a controller. For the time staying within a sector, the aircrafts position, altitude, and route, is tracked and partly annotated to the strip. When an airplane passes out of one controllers airspace and enters the space which is assigned to the control station of a colleague, he passes or throws the strip to him.

From a reductive or dualist point of view, the utility of paper strips is easily dismissed. Adopting a reductive or dualist stance reduces the role of paper strips to symbolic facts separated from their actual usage, or context. As the study illustrates:

1. A controller acts continously and deliberately out of himself, but not on the basis of calculative rationality. Thinking about rules for choosing goals, e.g., efficiency and safety, and reasons, e.g., weather conditions or traffic volume, for choosing possible actions like passing a strip distracts the continuous flow of activity. Conceptually, actions are not pre-determined (Indeterminism).

2. A controller doesn't have to step back and think about rules for authorizing routes. Instead, the physical layout of the strip provides a temporal and spatial proxy for managing his involvement and responsibility on the ground. The strip extends his cognitive capacity as a tactile and visual memory beyond objective time and space (Extension).

3. Picking up strips from the printer, placing strips on the tray, reordering strips as well as holding strips in hand adjusts mental load towards a flowing sense of owning the aircraft. Paper strips afford to act upon them by constituting a space of potentiality that is optimally adjusted (Potentiality).

4. There is a continuous checking of each aircraft on the radar and then on the strip. This checking is integrated into the controllers conceptual understanding of what it means to steer not a remote aircraft but his embodied and situated engagement. Due to the non-separability of controller and air traffic the combined state of both is best described as being-in-the-air (Non-separability).

Adopting a reductive or dualist perspective on work practices is certainly possible. However, it brings some problematic assumptions with regard to the relation between mind and matter. A quantum phenomenology, in particular the concept of dynamic co-emergence transcends reductive and dualist stances. It broadens the conceptual scope for more adequate and intelligible descriptions of work practices.

References

1. Dreyfus, H.: Being-in-the-world: A commentary on Heidegger's Being and Time, Division I. MIT Press, Cambridge (1991)
2. Heidegger, M.: Being and Time. Blackwell Publishing, Malden (1962)
3. Dennett, D.: Darwin's dangerous idea: Evolution and the meanings of life. Simon & Schuster, New York (1995)
4. Searle, J.: Mind: A Brief Introduction. Oxford University Press, Oxford (2004)
5. Chalmers, D.: Facing up to the problem of consciousness. In: Explaining Consciousness: The Hard Problem, pp. 9–32. MIT Press, Cambridge (1997)
6. Skinner, B., Frederic, B.: Science and Human Behavior. Free Press, New York (1965)
7. Kosslyn, S.: Image and Brain. MIT Press, Cambridge (1994)
8. Kim, J.: Emergence: Core ideas and issues. Synthese 151, 547–559 (2006)
9. Hanna, R., Thompson, E.: The mind-body-body problem. Theoria et Historia Scientiarum: International Journal for Interdisciplinary Studies 7, 24–44 (2003)
10. Atmanspacher, H.: Contextual emergence from physics to cognitive neuroscience. Journal of Consciousness Studies 14, 18 (2007)
11. Bitbol, M.: Ontology, matter and emergence. Phenomenology and the Cognitive Sciences 6, 293–307 (2007)
12. Thompson, E.: Mind in Life - Biology, Phenomenology and the Sciences of Mind. Harvard University Press, Cambridge (2007)
13. Filk, T., von Müller, A.: Quantum Physics and Consciousness: The Quest for a Common Conceptual Foundation. Mind and Matter 7(1), 59–79 (2009)
14. Mackay, W.: Is Paper Safer? The Role of Paper Flight Strips in Air Traffic Control. ACM Transactions on Human-Computer Interaction 6(4), 311–340 (1999)

Envisioning Dynamic Quantum Clustering in Information Retrieval

Emanuele Di Buccio and Giorgio Maria Di Nunzio

Department of Information Engineering – University of Padua
Via Gradenigo, 6/a – 35131 Padua – Italy
{emanuele.dibuccio,giorgiomaria.dinunzio}@unipd.it

Abstract. Dynamic Quantum Clustering is a recent clustering technique which makes use of Parzen window estimator to construct a potential function whose minima are related to the clusters to be found. The dynamic of the system is computed by means of the Schrödinger differential equation. In this paper, we apply this technique in the context of Information Retrieval to explore its performance in terms of the quality of clusters and the efficiency of the computation. In particular, we want to analyze the clusters produced by using datasets of relevant and non-relevant documents given a topic.

1 Introduction

Clustering is an unsupervised learning method for automatically organizing a large data collection by partition a set data, so the objects in the same cluster are more similar to one another than to objects in other clusters. The goal of clustering is to separate a finite unlabeled data set into a finite and discrete set of natural, hidden data structures, rather than provide an accurate characterization of unobserved samples generated from the same probability distribution [1]. This problem is inherently ill-posed in the sense that any given set of objects can be clustered in different ways with no clear criterion for preferring one clustering over another. This makes clustering performance very difficult to evaluate, since we have no targets and usually do not know a priori what groupings of the data are best. Despite this, the success of clustering methods as tools for describing the structure of data in a way that people can understand has been recognized in various areas of computer science [2].

In this paper we study a possible application of a recently proposed clustering method, known as Dynamic Quantum Clustering (DQC) [3], to the field of Information Retrieval (IR). We investigate the feasibility of the application of this method to the problem of document clustering. In particular, we want to tackle the following problems: how feature reduction impacts the quality of the clusters; how the reduction of the matrix in terms of selection of principal components affects the effectiveness of the method.

The paper is organized as follows: Section 2 discusses the problem of document clustering and the problem of textual clustering in IR; Section 3 presents the DQC method. Section 4 presents the experimental methodology and the experiments carried out. In Section 5 we make some final remarks.

D. Song et al. (Eds.): QI 2011, LNCS 7052, pp. 211–216, 2011.

2   Document Clustering

Document clustering has become an increasingly important task for analyzing huge numbers of documents. One of the challenging aspect is to organize the documents in a way that results in better search without introducing much extra cost and complexity. For a review on document clustering method, we suggest [4]. Typically, an IR system returns, as a response to a users query, a ranked list of documents. Nevertheless, several alternative organizations of the results have been investigated over recent years, most of them relying on document clustering [5], to reduce the users cognitive efforts. For example, query-specific clustering addresses the categorization of the first documents retrieved by an initial IR system with the aim of guiding the user in his search [6].

Initially, document clustering was suggested both for reasons of efficiency, since matching against centroids might be more efficient than matching against the entire collection, and as a way to categorize or classify documents [5]. Salton did early experimentation with document clustering, viewing clustering as classi-fication of documents in a manner similar to bibliographic subject headings. He wrote [7] "In a traditional library environment, answers to information retrieval requests are not usually obtained by conducting a search throughout an entire document collection. Instead, the items are classified first into subject areas, and a search is restricted to items within a few chosen subject classes. The same de-vice can also be used in a mechanized system by constructing groups of related documents and confining the search to certain groups only."

A basic assumption in retrieval systems is that documents relevant to a request are separated from those which are not relevant, i.e. the relevant documents are more like one another than they are like non-relevant documents. The cluster hypothesis [8] is fundamental to the issue of improved effectiveness. This hy-pothesis states that relevant documents tend to be more similar to each other than to non-relevant documents and therefore to appear in the same clusters. If the cluster hypothesis holds for a particular document collection, then relevant documents will be well separated from non-relevant ones. A relevant document may be ranked low in a best-match search because it may lack some of the query terms. In a clustered collection, this relevant document may be clustered together with other relevant items that do have the required terms and could therefore be retrieved through a clustered search. In this way, document clustering offers an alternative for file organization to that of best-match retrieval.

3   Dynamic Quantum Clustering

In DQC the problem of clustering data is mapped into a problem of quantum mechanics. The advantage of this mapping is that the techniques and concepts of quantum theory can be applied to reveal the clusters themselves. The basic idea is the following: each data point (i.e. a document) is associated with a particle that is part of a quantum system and has a specific field around its location. The state of the system is fully specified by a function $\psi(\mathbf{x}, t)$ that depends on the coordinates $\mathbf{x}$ of the particle in a specific point in time t. The probability that a particle lies in a volume of space $d\mathbf{x}$ located at $\mathbf{x}$ at time t is $|\psi(\mathbf{x}, t)|^2 d\mathbf{x}$ [9]. If the system is composed by N particles, the activation field in a location $\mathbf{x}$ is calculated by:

$$\psi(\mathbf{x}) = \sum_{j=1}^{N} e^{-\frac{\mathbf{x} - \mathbf{x}_j}{2\sigma^2}} , \tag{1}$$

where σ is a scale parameter.

Equation 1 is also known as Parzen window estimator (or kernel density estimator) which is a way of estimating the probability density of a random variable. In those regions of space where the data is denser, the Parzen window estimator would have relative maxima. The link between clustering and Parzen window estimator is the following: each local maximum can be seen as the centre of a cluster and the region around each maximum as the region belonging to that cluster. The drawback of this estimator is that it depends sensitively on the choice of σ: for small values of σ, too many local maxima and very small clusters are obtained; if σ is too large, the maxima are too smooth and no distinct clusters can be found.

Instead of using the Parzen window estimator directly, DQC uses it to construct a function whose minima are related to the clusters found by the estimator. The intuition is based upon the fact that in the quantum problem local maxima in the quantum state function (i.e. Equation 1) correspond to local minima in the potential function of the Schrödinger equation. DQC identifies these local minima by letting the particles of the quantum system to "roll down" into the local minima of the potential function. This is performed by defining the evolution of each state the system to be

$$\psi(\mathbf{x}, t) = e^{-iHt}\psi(\mathbf{x}) \tag{2}$$

where H is the Hamiltonian operator, i the imaginary unit, and e^{-iHt} is the time development operator. This time evolved state is the solution to the time-dependent Schrödinger equation:

$$i\frac{\partial \psi(\mathbf{x}, t)}{\partial t} = H\psi(\mathbf{x}, t) \equiv \left(-\frac{\hbar^2}{2m}\nabla^2 + V(\mathbf{x})\right)\psi(\mathbf{x}, t) , \tag{3}$$

where $-(\hbar^2/2m)\nabla^2$ is the kinetic energy operator, $V(\mathbf{x})$ the time-independent potential energy at position $\mathbf{x}$. The mass of the particle m is usually set equal to $1/\sigma^2$ and the reduced Planck constant $\hbar$ is absorbed by σ.

This apparently difficult problem of solving the time-dependent Schrödinger equation is reduced to the computation of simple closed form expressions followed by numerical evolution in the truncated Hilbert space, as explained in [3]. This solution reduces the problem to dealing with matrices whose size is determined by the number of data points and not by the dimension of features (i.e., the number of features associated with each document). Even in the case of a large number of points, there are considerations linked to the quantum theory that can help for dealing with that situation too (this problem is not tackled in this paper).

4   Experiments on DQC in Information Retrieval

The experiments were carried out on the TREC 2001 Web Track test collection which is constituted by a corpus of web pages, a set of fifty topics[1], and relevance assessments manually provided by human assessors on a set of documents in the corpus for a given topic. Experiments reported in the following are based only on two topics, 501 and 502; for each topic the following steps were performed:

1. consider the set $D_{J,q}$ of documents judged for the considered topic q;
2. select k terms to represent the documents; the selected terms are h terms extracted from the topic title and $k - h$ terms extracted from the documents in $D_{J,q}$ — stop words are not considered as candidate terms for selection;
3. prepare a term-by-document matrix $A \in \mathbb{R}^{k \times |D_{J,q}|}$ where the element $A_{j,i}$ is the weight $w_{i,j}$ of the term j in the document i;
4. apply Singular Value Decomposition (SVD) to A, thus decomposing A as $A = U \Sigma V^T$ and consider the first k' columns of the matrix V;
5. apply DQC to the matrix V^T.

The COMPACT software[2] was adapted in order to implement the above methodology steps. Step 2 aims at reducing the number of terms used to represent the documents, thus reducing the dimensionality of the matrix A. The number of distinct terms in $D_{J,q}$ is indeed over hundred thousand; term selection is needed because over a certain threshold for k, the computation becomes unfeasible. The experiments were performed by varying the value of $\sigma \in [0.01, 1]$, and the term selection strategy, specifically (i) the number of terms $k \in \{10, 100, 1000\}$, (ii) the set from which terms are extracted, and (iii) the weight for term ordering — only the $k - h$ terms with highest weight were retained. Terms were extracted from: a) the set $D_{J,q}$, b) only from the subset of documents judged as relevant, $D_R \subseteq D_{J,q}$, c) approximatively the same number of terms from the relevant document set D_R and the non relevant document set $D_{J,q} \setminus D_R$. Weights for term selection were computed by the Document Frequency (DF), the Inverse Document Frequency (IDF), and the RSJ term weighting [10].

The results obtained for the two topics when varying the source for terms were comparable; moreover, varying σ had no effect on the results. Table 1 reports the average values for diverse effectiveness measures computed over all three sources for terms and term weighting strategies for different numbers of principal components selected among those obtained by SVD. The results show a positive correlation between the adopted number of components and precision ($corr = 0.346$), a negative correlation with recall ($corr = -0.362$). The same result is also shown in Figure 1 with k' larger than 10. The results show that DQC clustering can benefit from a small number of components; it is possible to investigate the best trade-off between recall and precision by varying k'.

---

[1] A topic expresses a user information need; queries are derived from topic descriptions.
[2] http://www.protonet.cs.huji.ac.il/compact/

Table 1. Number of true positive, false positive, and values of recall and precision for different values of the adopted number of components, k'. Values are the mean computed over all the term selection strategies described in Section 4.

k'	True Positive	False Positive	Recall	Precision
2	50.590	1020.600	0.755	0.053
4	7.406	110.300	0.115	0.217
6	5.375	54.100	0.084	0.272
8	4.538	37.590	0.070	0.287
10	3.237	21.920	0.050	0.323

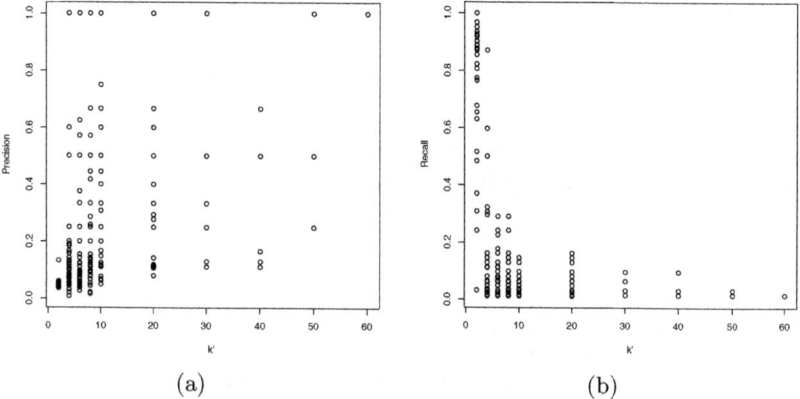

(a)   (b)

Fig. 1. The figures depict the relationship between k' and precision (Fig. 1a) and recall (Fig. 1b). Values refer to the results obtained for topics 501 and 502, all the term selection strategies and the sets of documents used as source for term selection. Similar trends have been observed when using only relevant documents as source for terms.

5 Final Remarks

In this paper we presented a study on a possible application of DQC to the IR field. In [3], the authors experiment on datasets with a limited number of features and therefore with spaces which are intrinsically dense. Our experiments considered datasets which have hundreds of thousands of features and are very sparse. The aim was to analyze the behavior of the DQC in these situations and how the selection of features and the principal components affect the clustering.

We tested the DQC on a standard test collection of IR and we made the following considerations: i) in order to reduce the initial space of features, different approaches to select the first k features were tested. There was no significant difference between the approaches and the number of features can be drastically reduced to a few tens without affecting the performances; ii)given the sparsity of the space, the analysis of the principal components showed that one can truncate the matrix of the SVD composition to the first ten or twenty values without affecting the performances of the clustering; adding more components does affect the effectiveness negatively; the truncation of the matrix is correlated positively with the precision of the clusters (less number of false positives) and negatively with the recall of correct documents (less number of true positives); iii) given the sparsity of the space, it is still not clear how to take advantage of the graphical inspection of the DQC. The resulting plot does not present the "roundness" given by the bi-modal, gaussian distribution of the datasets shown in [11,3]. This problem may be even more general in the sense that different underlying probability distribution may have a significant impact on the DQC effectiveness.

More recent experiments have confirmed that the results obtained for the different term selection strategies adopted showed that the only significant difference is between recall values obtained when using DF and IDF. DQC can achieve a 100% precision using a relative small number k' of principal components and for almost all the considered topics.

Acknowledgments. This work has been supported by the PROMISE network of excellence (contract n. 258191) project and by the QONTEXT project under grant agreement N. 247590 (FP7/2007-2013).

References

1. Xu, R., Ii: Survey of clustering algorithms. IEEE Transactions on Neural Networks 16(3), 645–678 (2005)
2. Becker, B., Kohavi, R., Sommerfield, D.: Visualizing the simple Bayesian classifier. In: Information Visualization in Data Mining and Knowledge Discovery, pp. 237–249 (2001)
3. Weinstein, M., Horn, D.: Dynamic quantum clustering: A method for visual exploration of structures in data. Phys. Rev. E 80(6), 066117 (2009)
4. Premalatha, K., Natarajan, A.M.: A literature review on document clustering. Information Technology Journal 9(5), 993–1002 (2010)
5. Hearst, M.A., Pedersen, J.O.: Reexamining the cluster hypothesis: scatter/gather on retrieval results. In: Proceedings of SIGIR 1996, pp. 76–84. ACM, New York (1996)
6. Lamprier, S., Amghar, T., Saubion, F., Levrat, B.: Traveling among clusters: a way to reconsider the benefits of the cluster hypothesis. In: Proceedings of SAC 2010, pp. 1774–1780. ACM, New York (2010)
7. Salton, G.: The SMART Retrieval System—Experiments in Automatic Document Processing. Prentice-Hall, Inc., Upper Saddle River (1971)
8. Jardine, N., van Rijsbergen, C.J.: The use of hierarchic clustering in information retrieval. Information Storage and Retrieval 7(5), 217–240 (1971)
9. Nasios, N., Bors, A.G.: Kernel-based classification using quantum mechanics. Pattern Recognition 40, 875–889 (2007)
10. Robertson, S.E., Jones, K.S.: Relevance weighting of search terms. Journal of the American Society for Information Science 27(3), 129–146 (1976)
11. Horn, D., Gottlieb, A.: Algorithm for data clustering in pattern recognition problems based on quantum mechanics. Phys. Rev. Lett. 88(1), 018702 (2001)

Contextual Image Annotation via Projection and Quantum Theory Inspired Measurement for Integration of Text and Visual Features

Leszek Kaliciak[1], Jun Wang[1], Dawei Song[1], Peng Zhang[1], and Yuexian Hou[2]

[1] The Robert Gordon University, Aberdeen, UK
[2] Tianjin University, Tianjin, China
{l.kaliciak,j.wang3,d.song,p.zhang1}@rgu.ac.uk; yxhou@tju.edu.cn

Abstract. Multimedia information retrieval suffers from the semantic gap, a difference between human perception and machine representation of images. In order to reduce the gap, a quantum theory inspired theoretical framework for integration of text and visual features has been proposed. This article is a follow-up work on this model. Previously, two relatively straightforward statistical approaches for making associations between dimensions of both feature spaces were employed, but with unsatisfactory results. In this paper, we propose to alleviate the problem regarding unannotated images by projecting them onto subspaces representing visual context and by incorporating a quantum-like measurement. The proposed principled approach extends the traditional vector space model (VSM) and seamlessly integrates with the tensor-based framework. Here, we experimentally test the novel association methods in a small-scale experiment.

Keywords: multimedia retrieval, quantum theory, image annotation, tensor product.

1 Introduction and Related Work

Despite the recent advancements in the field, multimedia information retrieval faces challenges. Most of them arise from the lack of a principled framework and the limitations of a widely used traditional vector space model (VSM), which finds it difficult to capture the inherent dependecies between entities (e.g. visual terms-textual description) and the contextual factors influencing the retrieval effectiveness.

Here, we are focusing on image retrieval. This area of research incorporates techniques from fields such as statistics, pattern recognition, signal processing, and computer vision, to analyze the content of images. Image retrieval also utilizes metadata information, e.g. tags and textual descriptions. The image content is usually represented as multidimensional vectors, which try to capture colour, shape or texture global or local (e.g. segmentation, "bag of features") properties.

It was experimentally proven (the annual imageCLEF competition results, for example) that a combination of textual and visual representations can improve the retrieval performance. However, most methods that utilize the combined information treat both types of features as separate systems. These approaches disregard the information about the inherent correlations between the different features' dimensions. Many methods

D. Song et al. (Eds.): QI 2011, LNCS 7052, pp. 217–222, 2011.

simply concatenate the representations or combine the scores ([7–9]). Others perform retrieval by text to pre-filter the images and then re-rank results by image content ([5]), or vice versa ([6]).

This paper is a continuation of the previous work on a unified framework [16], which incorporates a tensor product of textual and visual features. The tensor model requires that all images have textual annotations. Image annotation is a broad research area, therefore it would be difficult to refer to all interesting papers. In general, we can classify image annotation techniques into three groups (see [1]): recognition as translation, statistical models and combined approaches.

The first category of image annotation models may be compared to machine translation. Models try to predict one representation given another. Thus, [1] first performs image segmentation and then classifies the regions into corresponding "blobs" by utilizing k-means clustering. Next, the corresponding word for each blob is found by choosing the word with the highest probability computed by Expectation Maximization algorithm. However, due to the segmentation process, this approach can be computationally expensive, and the segmentation techniques do not always perform well.

Some methods utilize information about the correlations between so-called "visual words" (for more information about "bag of visual words" approach the reader is referred to [2]) and try to group semantically similar visual words' together. Such subsets of visual words can then be associated with textual terms. These approaches usually consider only co-occurrences at the local level and are computationally expensive and not scalable. Thus, Jamieson et al. [3] propose to group features that exist within a local neighbourhood, claiming that arrangements or structures of local features are more discriminative. Such groups of visual words are then associated with annotation words.

Approaches that belong to the second category of image annotation models, usually cluster image representations and text. In this way a joint probability distribution may be generated that link images and words. Finally, the labels for images that have high posterior probability may be predicted. For instance, [4] exploit statistical relationships between images and words without recognizing individual objects in images. This real-time annotation method, according to authors, can provide more than 98% images with at least one correct annotation out of the top 15 selected words. However, the high number of labels assigned to the given image may introduce a lot of noise in the form of, for example, contradictory meaning.

In this paper, we present and test two novel approaches for image annotation, which can be seamlessly integrated into the tensor-based framework. We also experiment with mid-level semantic content-based image representations based on the "bag of visual words" model. The first proposed method projects the unannotated image onto the subspace generated by subsets of training images. We calculate the probability of an image being generated by the contextual factors related to the same topic. In this way, we can capture the visual contextual properties of images, taking advantage of this extended vector space model framework. The other method introduced in this paper, performs quantum-like measurement on the density matrix of the unannotated image, with respect to the density matrix representing the probability distribution obtained from the subset of training images (containing given tags). These approaches can be seamlessly integrated into the unified framework for image retrieval [16].

2   Projection-Based Approach to Associating Textual and Visual Features

Many images do not have textual labels. Therefore, in order to prepare the data for the quantum-like measurement in the tensor space, we need to associate textual terms with images.

The idea behind the projection-based method is that dimensions of context define subspaces to which vectors of the information objects are projected ([12], see also [13]). Thus, we first build a density matrix[1] from the subsets of images containing the textual term t_i. This matrix represents a probability distribiution and incorporates information about the occurrence of some contextual factors (corresponding to basis vectors). It can be characterized in terms of co-occurrences between visual terms (e.g. visual words). Let y_i denotes the vector representation of the i-th image. Then the co-occurrence matrix A can be computed as

$$A = \sum_i |y_i\rangle\langle y_i| \tag{1}$$

Here, we assume that the correlations at the image-level may be stronger that the correlations based on the proximity between visual terms (instances of visual words are considered correlated if they appear together within a certain neighbourhood). An image may contain correlated terms (pixels, visual words) not because of their proximity, but because they refer to the same topic (image represents the context). The fore mentioned assumption was inspired by [14], where the page-based (text) correlations performed best. We will get back to discussing this problem later.

The symmetric correlation matrix A can then be decomposed to estimate the basis, which would represent the "relevance" context:

$$A = U \cdot F \cdot U^T = \sum_i f_i |u_i\rangle\langle u_i| \tag{2}$$

where U is a unitary, orthogonal matrix, f_i is an element of F and u_i are eigenvectors of A. Vectors u_i form an orthogonal basis of the subspace (as projector) representing the influence of each contextual factor. The projector onto this subspace (denoted as B) is equal to $P_B = \sum_i |u_i\rangle\langle u_i|$. $P(B)$ can be considered as the semantic subspace characterizing the term t_i. Now, each unannotated image d_i can be projected onto this subspace, and the probability of relevance context of d_i may be calculated as

$$Pr\,[L(B)|L(d_i)] = \langle d_i|P_B|d_i\rangle \tag{3}$$

where $L(d_i)$ denotes a subspace generated by d_i. Thus, the images are annotated with respect to the probability that they were generated by a context represented by P_B. The unannotated image can then be associated with a textual term corresponding to the semantic subspace with the highest probability of projection.

---

[1] The co-occurrence matrix is Hermitian and can be constructed in such a way that the trace would be unitary. Therefore the density and co-occurrence matrix will be used interchangeably in this paper.

3   Quantum Measurement-Based Approach to Associating Textual and Visual Features

Here, we introduce a variation of the projection method based on the quantum measurement. The proposed approach performs quantum like measurement on the density matrix A representing the probability distribution obtained from the subset of training images (containing given tags), and the density matrix D of an unannotated image d_i. Therefore

$$P_i = tr(D_i \cdot A) \tag{4}$$

where $D_i = |d_i\rangle\langle d_i|$.

4   Experimental Settings

In this paper we solely test the subspace-based auto-annotation methods. We manually choose a few terms and construct a subspace for each term. The projection of the unannotated image to the semantic subspace can be utilized to decide whether the image is about the term. We experiment on ImageCLEF 2007 data collection.

The terms are selected from the query text, some of which have explicit visual characteristics (e.g. sea), while others do not have general visual characteristics (e.g. california).

The measurement operator is constructed from all the images with relevant content. To simplify the experiment, we look at the available ground-truth data and choose 5 images that are specifically about the term, and construct the correlation matrix from term-document matrix M.

We manually select 10 relevant images belonging to each topic and 60 irrelevant images to investigate how the sub-space can distinguish the relevant from irrelevant images. The visual features we choose are: global colour histogram in HSV colour space, and local feature based on the bag of visual words approach. The latter consists of image sampling (random, dense sampling), description of local patches (three colour moments), quantization of descriptors (k-means) and generation of histograms of visual words counts.

The auto-annotation methods utilized in the experiments for the comparison are: projection (Eq. 3), quantum measurement (Eq. 4) and distance based. The latter clusters the training images containing given tags and the distance between cluster centroids and the unannotated image is used as the score for image - text association.

5   Results and Analysis

If a measurement operator can filter the relevant images with success, then this operator can be used to associate the text with visual features. The test results are shown in the Table 1. We can observe that cluster distance based measurement outperforms the other two. Here, *localdense* denotes local feature with dense sampling, *localrand* denotes local feature with random sampling, and *histHSV* is a colour histogram in HSV colour space.

Table 1. Accuracy of different measurements on various visual features. Here, q denotes quantum-like measurement, and p and d correspond to projection and distance based measurements respectively; the values in the table correspond to the number of positive associations at different precision levels.

	localdense						localrand						histHSV					
	q		p		d		q		p		d		q		p		d	
	p5	p10	p5	p10	p5	p10	p5	p10	p5	p10	p5	p10	p5	p10	p5	p10	p5	p10
mountain	1	2	1	2	1	2	0	1	0	1	1	1	1	2	0	1	1	2
sea	2	6	4	6	5	6	4	6	4	6	2	4	4	8	4	8	3	4
straight	0	1	0	1	2	4	1	2	1	3	4	5	0	2	0	1	3	4
black_white	4	9	4	8	5	9	4	8	3	8	5	9	5	9	5	9	5	8
girl	0	0	0	0	3	3	0	0	0	0	3	4	1	1	0	1	1	2
california	1	1	0	3	2	4	2	2	0	1	2	4	0	1	0	1	1	1

The observation is out of our expectations, as the subspace based measurement is supposed to capture the relevance of context as well as the latent information. This may be due to the small-scale experiment that was performed.

These results may be also related to our assumption that the correlations at the image-level may be stronger that the correlations based on the proximity between visual terms (pixels, local patches). An image may contain correlated terms (pixels, visual words) not because of their proximity, but because they refer to the same topic (context represented by image). We were inspired by [14], where the page-based correlations (text) performed better than proximity based ones. We were aware, however, that this does not have to be transferable to image retrieval. Further experiments will verify this hypothesis.

6   Conclusion and Future Work

In this paper, we describe and test two novel approaches for making associations between tags and images. We also experiment with mid-level semantic image representations based on the "bag of visual words" model. This is a follow-up work on the tensor-based unified image retrieval framework. In order to prepare the data for the quantum measurement in the tensor space, we need to alleviate the problem regarding the unannotated images. The first proposed approach projects the unannotated images onto the subspaces generated by subsets of training images (containing given textual terms). We calculate the probability of an image being generated by the contextual factors related to the same topic. In this way, we should be able to capture the visual contextual properties of images, taking advantage of this extended vector space model framework. The other method introduced in this paper, performs quantum like measurement on the density matrix of unannotated image, with respect to the density matrix representing the probability distribution obtained from the subset of training images. These approaches can be seamlessly integrated into the unified framework for image retrieval ([16]).

The experimental results show that the standard approach based on clustering works better than other methods. This may be due to the small-scale experiments conducted.

Another reason for these suprising results may be related to the assumption we made, that the correlations at the image-level may be stronger that the correlations based on the proximity between visual terms (pixels, local patches). Recent works build the correlations based on the proximity between image patches to capture the spatial information, as researchers believe that the relative distance between them is important. Thus, we need to test this alternative method for correlation matrix generation and perform large scale experiments.

References

1. Duygulu, P., Barnard, K., de Freitas, J.F.G., Forsyth, D.: Object Recognition as Machine Translation: Learning a Lexicon for a Fixed Image Vocabulary. In: Heyden, A., Sparr, G., Nielsen, M., Johansen, P. (eds.) ECCV 2002. LNCS, vol. 2353, pp. 349–354. Springer, Heidelberg (2002)
2. Yang, J., Jiang, Y.G., Hauptmann, A.G., Ngo, C.W.: Evaluating Bag-of-Visual-Words Representations in Scene Classification. In: Proc. of the Int. Workshop on Multimedia IR, vol. 206 (2007)
3. Jamieson, M., Dickinson, S., Stevenson, S., Wachsmuth, S.: Using Language to Drive the Perceptual Grouping of Local Image Features. In: IEEE Comp. Society Conference on Comp. Vision and Pattern Rec., vol. 2, pp. 2102–2109 (2006)
4. Li, J., Wang, J.Z.: Real-Time Computerized Annotation of Pictures. IEEE Tran. on Pattern Anal. and Machine Int. 30, 985–1002 (2008)
5. Yanai, K.: Generic Image Classification Using Visual Knowledge on the Web. In: Proc. of the 11-th ACM Int. Conf. on Multimedia, pp. 167–176 (2003)
6. Tjondronegoro, D., Zhang, J., Gu, J., Nguyen, A., Geva, S.: Integrating Text Retrieval and Image Retrieval in XML Document Searching. In: Advances in XML Inf. Retr. and Evaluation (2005)
7. Rahman, M.M., Bhattacharya, P., Desai, B.C.: A Unified Image Retrieval Framework on Local Visual and Semantic Concept-Based Feature Spaces. J. Visual Communication and Image Representation 20, 450–462 (2009)
8. Simpson, M., Rahaman, M.M.: Text and Content Based Approaches to Image Retrieval for the ImageClef2009 Medical Retrieval Track. In: Working Notes for the CLEF 2009 Workshop (2009)
9. Min, P., Kazhdan, M., Funkhouser, T.: A comparison of text and shape matching for retrieval of online 3D models. In: Heery, R., Lyon, L. (eds.) ECDL 2004. LNCS, vol. 3232, pp. 209–220. Springer, Heidelberg (2004)
10. van Rijsbergen, C.J.: The Geometry of Information Retrieval. Cambridge University Press, Cambridge (2004)
11. Griffiths, R.B.: Consistent Quantum Theory. Cambridge University Press, Cambridge (2003)
12. Melucci, M.: Context Modeling and Discovery Using Vector Space Bases. In: Proc. of the ACM Conf. on Inf. and Knowledge Management, pp. 808–815 (2005)
13. Di Buccio, E., Melucci, M., Song, D.: Towards Predicting Relevance Using a Quantum-Like Framework. In: The 33rd European Conference on IR, pp. 19–21 (2011)
14. Biancalana, C., Lapolla, A., Micarelli, A.: Personalized web search using correlation matrix for query expansion. In: Cordeiro, J., Hammoudi, S., Filipe, J. (eds.) Web Information Systems and Technologies. LNBIP, vol. 18, pp. 186–198. Springer, Heidelberg (2009)
15. Aharonov, Y., Albert, D.Z., Au, C.K.: New Interpretation of the Scalar Product in Hilbert Space. Phys. Rev. Lett. 47, 1029–1031 (1981)
16. Wang, J., Song, D., Kaliciak, L.: Tensor Product of Correlated Text and Visual Features: A Quantum Theory Inspired Image Retrieval Framework. In: AAAI-Fall 2010 Symp. on Quant. Inf. for Cognitive, Social, and Semantic Processes, pp. 109–116 (2010)

MPEG-7 Features in Hilbert Spaces: Querying Similar Images with Linear Superpositions

Elisa Maria Todarello[1], Walter Allasia[1], and Mario Stroppiana[2]

[1] Eurixgroup
[2] RAI CRIT

Abstract. This work explores the analogies between an Information Retrieval (IR) task and the process of measuring an observable quantity in Quantum Mechanics (QM) applied to digital images represented by MPEG-7 Visual Descriptors. Clusters of images are described as superpositions of vectors, taking into account the distribution of the feature values of all the members. Similarity scores are computed making use of the geometric structure of Hilbert spaces with part of the rules of QM and used to compute cluster assignments. A software prototype has been developed to test the method.

Keywords: MPEG-7, image, Hilbert, information, retrieval, quantum, mechanics, superposition, similarity.

1 Introduction

We describe a method for the representation of digital images, clusters of images and queries as vectors of a Hilbert state space equipped with part of the rules of Quantum Mechanics (QM). Clusters are naturally represented as linear superpositions of the vectors members. The similarity between images (single or clusters) and queries is computed thanks to the geometrical structure of the space enabled by the definition of a scalar product. We also present a prototype software implementation of the method applied to sets of digital images having MPEG-7 Visual Descriptors. We ran a preliminary test evaluation of the developed software computing cluster assignments of sample query images.

Section 2 reports a quick overview of the works and theories analised for the development of this work. The proposed method is described in Section 3. The software implementation and test are described in Section 4. Conclusions and future work hints are contained in Section 5.

2 Related Work

The idea of an analogy between the elements of QM and IR was firstly theorised in Ref. [8]. In Ref. [8], the relevance of a document with respect to a query is associated to a Hermitian operator **R** on a Hilbert space where objects are represented as normalized vectors. By means of the Gleason's Theorem, a probability

D. Song et al. (Eds.): QI 2011, LNCS 7052, pp. 223–228, 2011.

measure is defined on each subspace. Ref. [8] explicitly states that the analogy is general enough to be applied to any kind of document (text, image, etc.).

After Ref. [8], several research groups started working in to apply the idea. An overview of the state of the art for QM and IR can be found in Ref. [7]. Some important works oriented to retrieval of text documentsare: Ref. [5], focused on the description of context in a IR task; Ref. [6], devoted to the modeling of user interaction. Ref. [9] proposes a technique to unify annotation-based and content-based retrieval of digital images, based on HSV features.

We propose an application of the analogy as well, in particular to the Content-Based IR (CBIR) of digital images, with the use of MPEG-7 Visual Descriptors (Ref. [3]) as visual features.

To create the method, we started from the QM formalism, thoroughly described in Ref. [2]. The state of a physical system is represented by a normalized *state vector*, $|\psi\rangle \in \mathcal{H}$, where $\mathcal{H}$ is the Hilbert *state space*. An observable quantity, $\mathcal{X}$, is represented in $\mathcal{H}$ by a Hermitian operator, $\mathbf{X}$. The eigenvalues of $\mathbf{X}$ are the possible results of a measurement of $\mathcal{X}$. Each eigenvalue is associated to a subspace of $\mathcal{H}$ through the eigenvalue equation of the operator. The (normalized) eigenvectors of $\mathbf{X}$ form an orthonormal basis for $\mathcal{H}$. This shows that $\mathcal{H}$ has one dimension for each possible value of $\mathcal{X}$.

The probability $\mathcal{P}_\psi(x_i)$ of getting the eigenvalue x_i as the result of a measurement of $\mathbf{X}$ on a system in the state $|\psi\rangle$ is given by the orthogonal projection of $|\psi\rangle$ onto the corresponding eigensubspace:

$$\mathcal{P}_\psi(x = i) = |\omega_i|^2 = \langle\psi|\mathbf{P}_i|\psi\rangle \ , \tag{1}$$

where $\mathbf{P}_i$ is the projector onto the eigensubspace associated with x_i and ω_i is the probability amplitude of getting x_i when a measurement of $\mathbf{X}$ is performed. Commutation rules have to be defined between operators. If the two operators commute, then it is possible to find a set of eigenvectors that solves both the eigenvalue equations and the operators form a Complete Set of Commuting Observables (CSCO).

3   Image Representation and Relevance Computation

We illustrate a method for the representation of images, clusters of images and queries by sample in a single Hilbert space built in analogy with the state space of QM. The similarity score between images is used for cluster assignment of the queries. This approach is well-suited for the representation of clusters as linear superpositions of vectors. The assignment of an image to a cluster is then naturally associated to the QM process of measuring an observable quantity.

We illustrate the mapping of the elements of QM to IR concepts through a simple example. Consider images characterised by one visual feature, $\mathcal{X}$, which can assume the quantized values 0,1,2 (the method can be applied to any feature).

1. $\mathcal{X}$ is represented as a Hermitian operator, $\mathbf{X}$ on the Hilbert space $\mathcal{H}$.
2. The eigenvalues of $\mathbf{X}$ coincide with the values the feature can assume, in our example $x_0 = 0, x_1 = 1, x_2 = 2$. In the example, we assume the eigenvalues are non-degenerate.
3. The state space $\mathcal{H}$ is the span of the eigenvectors of $\mathbf{X}$. Denoting the eigenvector associated to x_i as $|x_i\rangle$, an orthonormal basis for $\mathcal{H}$ is $\{|x_0\rangle, |x_1\rangle, |x_2\rangle\}$.
4. An image document d is associated to a vector $|d\rangle \in \mathcal{H}$, thus it can be expressed as a linear combination of the basis vectors. In the example, $|d\rangle = \sum_{i=0}^{2} \omega_i |x_i\rangle$, where the ω_is are the probability amplitudes.

The scalar product in $\mathcal{H}$ is defined in the usual way on the basis of the eigenvectors of the operator. All the considered vectors are normalized to 1. $\mathcal{H}$ has one dimension for each possible value of $\mathbf{X}$. An image document d having the value 0 for $\mathcal{X}$ will be represented as $|d\rangle = |0\rangle$. This representation is trivial for single images: the image vector always coincide with one of the eigenvectors of $\mathbf{X}$, i.e. it has a well-defined value for the feature $\mathcal{X}$.

According to the principle of superposition any linear combination of vector images must also represent a vector in $\mathcal{H}$. The introduction of clusters of images naturally provides a meaning for the QM principle of superposition in the IR analogy: a cluster is a vector in $\mathcal{H}$, represented as a linear superposition of the vector images belonging to the cluster. The probability amplitudes ω_i are defined as the square roots of the occurrence frequencies of each vector in the cluster. A clusters C including the elements $C = \{|d_i\rangle\}$ $i = 1, \dots, N$ is then described as:

$$|C\rangle = \sum_{i=1}^{N} \omega_i |d_i\rangle \tag{2}$$

$$\omega_i = \sqrt{\frac{\text{number of occurrences of } d_i}{\text{total number of vectors } C}} \quad \text{where} \quad \sum_i |\omega_i|^2 = 1 \; . \tag{3}$$

If $C = \{|0\rangle, |0\rangle, |1\rangle, |1\rangle, |2\rangle\}$, then $|C\rangle = \sqrt{\frac{2}{5}}|0\rangle + \sqrt{\frac{2}{5}}|1\rangle + \sqrt{\frac{1}{5}}|2\rangle$. This representation carries more information than the identification of a cluster with its barycenter (or centroid), since the cluster is associated to a probability distribution. In usual techniques for image clustering, an image is assigned to the cluster whose centroid is the closest to the image vector. In the case of a strongly scattered vector distribution in the cluster this could produce association errors.

5. The query q is an image to be assigned to a cluster. It is associated to the concept of state of a quantum system, because it induces a probability measure on the subspaces of the state space, as stated by the Gleason's Theorem, i.e. it assigns a probability of relevance to every image/cluster. It is denoted in $\mathcal{H}$ as $|q\rangle$.
6. The probability of relevance of an image/cluster C with respect to q is

$$\mathcal{P}_q(C) = \langle q|\mathbf{P}_C|q\rangle \tag{4}$$

where $\mathbf{P}_C$ is the orthogonal projection operator onto $|C\rangle$.

The assignment of q to a cluster is determined by computing the similarity between q and each available cluster vector and choosing the most similar one.

If there is more than one feature, the set of the corresponding operators form a CSCO. This limitation may be removed in the future.

Not all the Postulates of QM find a correspondence in this method: we didn't find a meaning for the existence of non-commuting operators and for the collapse of the state vector. Also, a Hamiltonian evolution of the system is not provided.

4   Implementation and Testing

We developed a software prototype that implements the method, applied to digital images characterized by the following MPEG-7 Visual Descriptors:

- the Scalable Color, with 64 Haar coefficients with 8 bitplanes;
- the Color Layout, with 6 coefficients for luminance and 3 coefficients for each chrominance;
- the Edge Histogram, with 5 types of edges in the 16 subdivision blocks of the image, resulting in 80 coefficients.

At the first stage, images were represented in a usual "metric space", where each dimension corresponds to a coefficient and the similarity between two images is in inverse proportion to the distance between the vectors. The vectors in the "metric" space were clustered using the tool Kmlocal (Ref. [4]), with the Hybrid implementation of the k-means algorithm. This tool outputs the cluster barycenter coordinates and the cluster assignment of each image in the data set. At the second stage, the "Hilbert space" was created following the model described in Section 3.

To test the application, we used a data set of 18461 images randomly selected from the CoPhIR collection (Ref. [1]), available with the MPEG-7 Visual Descriptors. The images were grouped into 1846 clusters. The number of clusters was chosen according to a rule of thumb. We chose 12 query images from the data set and assigned them to a cluster according to the two methods available in the implementation: the "metric method" (assignment computed at the time of cluster creation) and the "Hilbert method", for which we created the representation of the queries and the clusters in the Hilbert space, then scored each query against each cluster with the algorithm of Eq. 4. The query is assigned to the cluster that gets the highest score.

As an illustrating example, consider a cluster C containing 10 members. Assuming there is one coefficient, Scalable Color 0, that can assume only integers values from -19 to 20, the Hilbert space has 40 dimensions. Having $|C\rangle = \sqrt{\frac{2}{10}}|11\rangle + \sqrt{\frac{5}{10}}|12\rangle + \sqrt{\frac{3}{10}}|13\rangle$ means that in C there are 2 member images with Scalable Color 0 equal to 11, 5 members with Scalable Color 0 equal to 12, and 3 members with Scalable Color 0 equal to 13. The other basis vectors have a null coefficient.

Due to the fact that the data set didn't have a clear cluster structure, the clusters created were not easily identifiable with a specific content or subject. In some cases, it was not possible to evaluate some of the assignments because of the poor quality of the clusters. Moreover, the clusters were created for this testing, so the queries had no pre-defined assignment. This means that the results had to be evaluated visually, with a qualitative comparison of the two methods. Table 1 reports the results of this visual evaluation. Software time performance was not evaluated at this stage.

Table 1. Visual Evaluation comparing the query assignments with the "metric" and "Hilbert" methods. Query Ids are the file names in the CoPhIR collection.

N.	Query Id	Visual Evaluation
1	9656496	Clusters are visually similar
2	35404821	Same cluster
3	67528271	Assignment with Hilbert space method not visually good
4	24869474	Same cluster
5	67154639	A dark area in the lower part of the query image determines the assignment to a different cluster
6	67867803	Same cluster
7	5042589	Same cluster
8	5042740	Same cluster
9	67479156	The metric assignment is visually better
10	24567694	Same cluster
11	35685661	The clusterization was not good
12	41930413	The clusterization was not good

5   Conclusions and Future Work

The first contribution of this work is a method for the representation of images by means of their features in a Hilbert space $\mathcal{H}$ built in analogy with the QM state space: a feature is represented by Hermitian operators, whose eigenvalues are the possible values of the feature. The query is associated to the QM state vector. Clusters of images are documents in $\mathcal{H}$ represented by the linear combinations of the vectors making up the clusters. This provides a meaning for the principle of superposition. Images, clusters and queries are then represented uniformly as vectors of the same Hilbert space. The similarity between a query-image and any image/cluster in the data set is given by Eq. 4.

The second contribution is a software prototype application, implementing the method for images described by MPEG-7 Visual Descriptors. The software prototype creates the Hilbert space and computes the assignment of a query image to one of the clusters. The testing of the application provides a qualitative evaluation of this assignment in comparison with the usual barycenter-based method. Results are reposted in Table 1.

In future, the software application needs to be improved: the new method should be used for the creation of the clusters. Also, the use of a data with a more definite cluster structure and a ground truth data would allow a quantitative evaluation and the association of a *meaning* to the clusters. The parameters of the k-means algorithm should be fine-tuned. Finally, the queries should be images not used at the clusterization stage.

An aspect that needs further investigation is the application to image IR of the algebraic properties of incompatible observables. Incompatibility can arise if we consider more than one description criterion of the same image. Also, quantum contextuality can be used to model user interaction with the system.

A technique that can be used to model user interaction is pseudo-relevance feedback, to follow automatically the user's choices and interests. If a user makes a query to a system, the first k vectors of the result set can be represented as a cluster-superposition and the image query can be rotated in this k-subset, in order to reapply a further query, more precisely addressing the user needs.

Local features are arising such as the Scale Invariant Features Transformations. The MPEG standardization body is going to analyze and standardize local features as Compact Descriptors for Visual Search (CVDS group). These new descriptors can be used for a future version of the application. The presented method may also be used when different kind of documents are in the cluster, such as images and videos.

References

[1] Bolettieri, P., Esuli, A., Falchi, F., Lucchese, C., Perego, R., Piccioli, T., Rabitti, F.: CoPhIR: a Test Collection for Content-Based Image Retrieval. CoRR abs/0905.4627v2 (2009)

[2] Cohen-Tannoudji, C., Dui, B.: Quantum Mechanics. Wiley-Interscience, Hoboken (1991)

[3] JTC1/SC29/WG11, I.O.F.S.I.: Information Technology - Multimedia Content Description Interface Part 3: Visual, ISO/IEC 15938-3:2002 (2001)

[4] Kanungo: An Efficient K-Means Clustering Algorithm: Analysis and Implementation. IEEE Trans. Pattern Analysis and Machine Intelligence 24, 881–892 (2002)

[5] Melucci, M.: A Basis for Information Retrieval in Context. ACM Transaction on Information Systems 26(3), 41 pages (2008)

[6] Piwowarski, B., Lalmas, M.: Structured information retrieval and quantum theory. In: Bruza, P., Sofge, D., Lawless, W., van Rijsbergen, K., Klusch, M. (eds.) QI 2009. LNCS(LNAI), vol. 5494, pp. 289–298. Springer, Heidelberg (2009) (last visited: November 12, 2010)

[7] Song, D., Lalmas, M., van Rijsbergen, K., Frommholz, I., Piwowarski, B., Wang, J., Zhang, P., Zuccon, G., Bruza, P., Arafat, S., Azzopardi, L., Buccio, E.D., Huertas-Rosero, A., Hou, Y., Melucci, M., Rueger, S.: How Quantum Theory Is Developing the Field of Information Retrieval (2010)

[8] van Rijsbergen, C.J.: The Geometry of Information Retrieval. Cambridge University Press, Cambridge (2004)

[9] Wang, J., Song, D., Kaliciak, L.: Tensor Product of Correlated Textual and Visual Features: A Quantum Theory Inspired Image Retrieval Framework. In: AAAI Fall Symposium Series (2010)

Author Index

Aerts, Diederik   25, 95, 116, 192
Aerts, Sven   13
Allasia, Walter   223
Arafat, Sachi   161
Asano, Masanari   182
Atmanspacher, Harald   105, 128

Basieva, Irina   182
beim Graben, Peter   105
Broekaert, Jan   192
Bruza, Peter   149
Busemeyer, Jerome   71

Cohen, Trevor   48
Czachor, Marek   192

Darányi, Sándor   2, 60
D'Hooghe, Bart   95, 192
Di Buccio, Emanuele   211
Di Nunzio, Giorgio Maria   211

Economides, George   199
Eyjolfson, Mark   25

Filk, Thomas   105, 128
Flender, Christian   205
Fuchs, Christopher A.   1

Gabora, Liane   25
Galea, David   149
Grefenstette, Edward   35

Hou, Yuexian   217
Huertas-Rosero, Alvaro Francisco   138

Kaliciak, Leszek   217
Khrennikov, Andrei   182
Kitto, Kirsty   13, 149

Lambert-Mogiliansky, Ariane   71

McEvoy, Cathy   149

Nelson, Douglas   149

Ohya, Masanori   182

Rindflesch, Thomas C.   48

Sadrzadeh, Mehrnoosh   35
Schvaneveldt, Roger   48
Sitbon, Laurianne   13
Smith Jr., Charles E.   83
Song, Dawei   217
Sozzo, Sandro   95, 116
Stroppiana, Mario   223

Tanaka, Yoshiharu   182
Tarrataca, Luís   172
Todarello, Elisa Maria   223

van Rijsbergen, C.J.   138
Veloz, Tomas   25

Wang, Jun   217
Wichert, Andreas   172
Widdows, Dominic   48
Wittek, Peter   2, 60

Yamato, Ichiro   182

Zhang, Peng   217
Zorn, Christopher   83

GPSR Compliance

The European Union's (EU) General Product Safety Regulation (GPSR) is a set of rules that requires consumer products to be safe and our obligations to ensure this.

If you have any concerns about our products, you can contact us on ProductSafety@springernature.com

In case Publisher is established outside the EU, the EU authorized representative is:

Springer Nature Customer Service Center GmbH
Europaplatz 3
69115 Heidelberg, Germany

Batch number: 09490872

Printed by Printforce, the Netherlands